CW01498010

The Murex trunculus or banded rock murex

Excerpt From The Jewish Journal Book Award Announcement - 2013

The making of a memorable book requires the skills of an alchemist. Every author starts with the raw material of his or her own experience and expertise, but it can take a certain secret ingredient — passion, vision, inspiration — to transform the dross into gold. That is a fair description of what Baruch Sterman and Judy Taubes Sterman have accomplished in *The Rarest Blue: The Remarkable Story of an Ancient Color Lost to History and Rediscovered...*

Because the Stermans possess precisely that alchemical genius, the Jewish Journal Book Prize for 2013 is awarded to *The Rarest Blue*, the second-annual prize given in recognition of a book of exceptional interest, achievement and significance...

The Rarest Blue begins in distant antiquity, moves forward through two millennia of Jewish culture and history, and drills down deeply on the scientific enterprise of more recent times. The key to the riddle of *tekhelet* is a marine snail known as the *Murex trunculus*, whose entrails were used by Bible-era dye-makers to create the hyacinth blue that is mentioned some 50 times in the Tanakh...

The annual winner of the annual Jewish Journal Book Prize is selected by the Journal's book editor. The prize is not restricted to Jewish authors or books on Jewish subjects, but *The Rarest Blue* is an example of a book that is worthy of attention both for its literary merit and for its Jewish interest.

- Jonathan Kirsch, Jewish Journal

"An amazing tale of historical sleuthing, told by a true storyteller who manages to capture both the drama and the magic of the quest for Biblical blue."

> \- Simcha Jacobovici, Emmy award winning journalist and director, and New York Times Bestselling author

"The topic of techeles is one worthy of serious study by all who are committed to halacha, and the Sterman's book, *The Rarest Blue*, presents the subject, in all its many facets, in a comprehensive and engaging manner. It was a pleasure to read!"

> \- Rabbi Hershel Schachter, Rosh Yeshiva, Yeshiva University

"The Stermans' dedication is admirable and their research comprehensive ... Ambitious."

> \- Kirkus Reviews

"Expansive and fascinating ... Sterman tackles a broad range of topics, his emphasis on Jewish traditions and ancient practices keeps it centered and illuminating."

> \- Booklist

Praise for
The Rarest Blue

"A story of science and religion, of craft and history. *The Rarest Blue* is spellbinding, each page a revelation. In lovely, engaging prose, the Stermans reveal the rediscovery of snail indigo – a detective story with cultural origins and a spiritual ending. A wonderful book."
- Roald Hoffmann, winner of the Nobel Prize in Chemistry

"Following in the footsteps of my grandfather, the Stermans leave no stone unturned in uncovering the secrets of the ancient, long sought after, biblical blue. History, archeology, religion, chemistry, and biology are skillfully woven together in *The Rarest Blue*, a fascinating book that will appeal to a wide range of readers."
- Isaac Herzog, President of the State of Israel

"*The Rarest Blue* can be enjoyed as a mystery, a travelogue, an adventure story and a work of scholarship…The story ends on a note of triumph that can be understood variously as an affirmation of piety or as the success of a scientific enterprise, or perhaps both."
- Jonathan Kirsch, Jewish Journal

You'll love this Indiana Jones-style chemistry adventure: an enlightening hunt for the lost source of history's most valuable dye."
- Mental Floss Magazine, November 2012 Issue

"According to tradition, as described beautifully in *The Rarest Blue*, white symbolizes purity, and blue, the color of the heavens, represents holiness. The white combined with the blue tekhelet conveys the message that a mortal can indeed achieve a state of holiness, and the tekhelet string points the direction to a truly spiritual life."
- Rabbi Abraham J. Twerski, M.D

Contents

Prologue

THE SETTING SUN

*T*he dazzling Greek sun had arched past its midpoint in the sky – which meant I didn't have much time. Driving down the coastal highway from Corinth to Athens at about ten miles per hour over the speed limit, I was beginning to feel anxious. In a few hours I had to be back at my hotel. With the setting of the sun, the Sabbath would begin, and for the next twenty-four hours I couldn't do any kind of work, according to the broadest definition of the term. That included driving my car. If I miscalculated the remaining time, I'd have to pull over to the side of the road, leave the car, and walk the rest of the way.

It wouldn't be the first time, I thought.

The night before, I had flown to Athens on my way back home to Israel from a business trip in Germany. I was working on a new idea for a laser that could turn radio waves into light energy. My company wanted to make the laser very compact, and the problem that I hoped my German colleague could help solve was how to efficiently channel radio waves into the laser, instead of merely letting them disperse as useless radiation. In the quaint town of Aachen, snow had fallen on a frozen brook while I was thinking about laser energy, and now, dressed in a T-shirt, I was driving along the Mediterranean coast, stopping in one village after another with something quite different in mind.

I had pretty much given up hope of finding it. The morning had begun with a trip to a small hardware store in Athens to buy a few items: hammer, razor blade, half-gallon jar. The man behind the counter looked at me curiously. In a thick Greek accent that stretched his English to the breaking point, he asked: "What you want for? You fix?"

"No, I dye," I replied. Then, quickly realizing that he probably imagined some gruesome suicide scene, I touched a colorful spot on my shirt. "You know, make color."

He looked at me with his head turned a bit to the side, the expression on his face clearly reading *Crazy foreigners. I can never understand what they are talking about.*

I had studied the map before setting out that morning and decided to drive all the way to Corinth, then double back to Athens. On the way I would stop in the villages and inquire among the local fishermen. It probably wasn't the most efficient plan. I should have started close to Athens and traveled farther west if necessary. But sundown was a hard deadline, so it made more sense to head toward my hotel as the day progressed.

Narrow and winding, the "highway" to Corinth keeps the coast on one side and mountains on the other. The scenery was a tourist's delight: Small whitewashed villages checkered the landscape – Stikas, Kineta, Agioi Theodoroi. But I hadn't come to enjoy the scenery or visit the tourist sites, not even the stunning Temple of Apollo in Corinth.

I was on a mission.

In each village that I entered, I looked for the marketplace or the harbor and asked the locals the same one-word question, the full extent of my Greek: "*Porphyros?*" Sometimes they simply stared at me, but every once in a while came a voluble response accompanied by hand waving and pointing that directed me to the next village down the road.

How many times had I done this already, I wondered as I stopped in front of a small store in the seaside village of Pyrgos. Sitting outside, an old woman dressed in black was sewing a button on a shirt. I smiled my politest smile and asked my Greek word: "*Porphyros?*" She looked at me as if she didn't hear me, so I took out a business card and wrote on the back πορφυρος. My doctorate in physics was unexpectedly coming in handy. One thing a physicist knows is the Greek alphabet, since we use those letters as various symbols in equations.

The old woman looked closely at the letters, nodded a few times, then pointed without saying a word, waggling her fingers. She wasn't steering me to the next village, but rather to a path leading in the direction of the sea. Had she understood me correctly? Were my efforts at last paying off? I smiled again, by way of thanks, and hurried down toward the beach. Fishing boats, docked at a pier, were bobbing in the afternoon sun. On the pier itself, fresh looking fish were piled high on stands. One stand, however, was not filled with fish but rather – to my great delight – with snails.

Tense with anticipation, I picked up a snail and examined it. It was bigger than those I knew, but the shape was similar. Down the front of the shell ran a ridge leading to a wide oval opening, and on the inner side I could make out the characteristic white and brown bands. With increasing excitement I looked at the other snails, a few of them peeking out of their shells, their eyes dangling at the ends of long stems. Some shells were shut tight, but the hard covering with which the snails sealed themselves into their shells was the familiar brown color with a smooth texture like a fingernail. There could be no doubt. It was the *Murex trunculus*. It was what I had been looking for.

I almost didn't believe my luck; here were hundreds of the snails, a whole treasure trove. I held up seven fingers, pointed to the snails, and said to the boy selling them, "Kilos." He looked at me in disbelief. Seven kilos amounted to about five hundred snails, enough for a major banquet. But these were not for eating.

He hurriedly weighed them, put them into a few boxes, and helped me carry them to the seashore. There, on a strong, level rock, I went to work. I had to be fast and efficient if I was going to make it back to my hotel room in time.

One by one, I took each snail and positioned it so that the opening was flat against the rock. I aimed about two-thirds of the way down the back, and with one firm stroke of my hammer broke through the shell. Then with my pinky, I pushed the snail itself deeply into its shell until part of the soft body came squeezing through the hole I had broken. If I did it right – and I did nearly every time – there, oozing through the hole on the back of the snail's shell, appeared a yellow gland, about a quarter of an inch long and the width of a piece of spaghetti. With the razor blade and my thumb as a counter force, I sliced off the gland and dropped it into the jar. The broken snail itself I tossed into the ocean.

Pretty soon I had an audience – two audiences, actually. In the water, dozens of fish congregated to feast upon the discarded snails. On the shore, local fishermen and children gathered with great curiosity to see what this stranger was doing. The kids, dark haired and tanned, laughed merrily as they pointed at me, whispering "*trelos*" – not a compliment, as I later learned.[*] I must have looked quite ridiculous as I cracked open snail after snail and performed microsurgery on each one.

Meanwhile, the yellow glands that I had been placing in the half-full jar were rapidly turning a bluish purple, as were my own hands. I finished shell breaking, rubbed my hands in the seawater to wash off the grime and the smell – the stain on my fingers, I knew, would last for weeks – waved to the children and the fishermen, and returned to my car. I even managed to make it back to my hotel in Athens with enough time for a quick shower before the Sabbath began. I placed the jar outside my window, hoping that the cool air would help keep the contents fresh.

The next morning, when I opened the window, the rancid smell of rotting fish nearly knocked me over. Finding and procuring the snail glands had been difficult enough, but getting that jar safely through security and onto the plane was going to require divine intervention. That night I wrapped it in layer after layer of plastic bags, until the stench became little more than a faint odor, and hoped for the best.

In Israel my friends Eliyahu, Joel, and Ari met me at Ben Gurion International Airport. They had received my e-mail announcing the great find, and they were too excited to wait until I got home to see the treasure. I unwrapped the jar, and everyone stared in awe at the bluish-purple liquid. It was a solemn moment; we were all participating in a venture that had historical significance – and we knew it. We also knew that we would be able to

---

[*] *Τρελός* means crazy.

find as many of the *Murex trunculus* snails as we wanted and that we could produce all the dye that we would need.

The secret of marine snail dyes had been lost for 1,300 years, but we were about to restore the sacred, rarest blue.

THE
RAREST

Blue

I
THAT DYE OF DYES

*T*oday we live in a world of endless and vibrant color. Glittering neon lights, LEDs, laser and ink-jet printers, paints, cosmetics, fabrics – all bombard us with eye-popping hues wherever we turn. In the ancient world, however, this multicolored richness didn't exist. The palette of the artists and craftsmen of old consisted largely of natural earth tones – the browns, beiges, ochers, and blacks found most prevalently in the environment. Prehistoric cave paintings all appear in these shades because their creators used rocks and stones close at hand to make their art.

Along with rocks and minerals, indigenous plant life also offered a source of color for the ancient world. Long ago it was observed that many flowers, leaves, stems, tree barks, and even vegetable peels yield pigments when boiled, primarily in the yellow and brown families. These plant and rock derivatives, used by artisans to color various materials, were essentially stains rather than dyes, though. They didn't bind permanently to textiles the way that a true dye does, and given the right circumstances – heavy rain, for example – stains run. Colored fabrics and clothes were rare in those days, and people dressed primarily in the drab tan, beige, and brown hues of the sheep and goats whose wool they spun, or whose hides they processed and tanned to make their clothing. When the patriarch Jacob in the Bible gave his favorite son, Joseph, that famous, unique gift, the "coat of many colors," his other sons reacted with fratricidal jealousy.

Dyes that didn't run, fade, or wash out were precious. One of the first of these came from the insect kingdom with the discovery that tiny bugs such as the scale insect, *Kermes ilicis,* which lives on oak trees in the Mediterranean, could be dried and crushed to produce a colorfast crimson dye. The word *crimson,* in fact, derives from the word *kermes.*

Muted tones dominated the spectrum in antiquity; rich shades like purple and blue, which occur rarely in nature, were inaccessible to ancient dyers. That is, until the discovery that certain shellfish, under certain conditions, could produce these vibrant hues and, more importantly, in a dye form of the most excellent, enduring quality. This discovery, which probably occurred independently in different parts of the globe at different times, had far reaching ramifications. Its influence extended to many aspects of human life, leaving its mark on the social, political, and religious realms, as well as the economic, intellectual, and scientific. Once the blue and purple dyes burst onto the scene, the world was never the same. A new range of beautiful colors became available, and the demand for them proved overwhelming. For nearly two millennia, all along the coast of the Mediterranean, from Jaffa to Djerba, shellfish dyes ranked among the most desirable commodities of the ancient world – and among the costliest.

As the dye industry developed, great resources and efforts went into perfecting the laborious dyeing techniques. Dye houses flourished. The prized merchandise found its way everywhere, bringing unprecedented fortunes to entrepreneurs up and down the entire production and distribution chain. Battles were fought over control of the dyeing industry, and strict laws regulated the manufacture and use of the dyes. Those who had mastered the arcane and exacting processes – from catching the snails to the elaborate methods for effective dyeing of fabric – guarded the esoteric knowledge with their lives.

But then this vast and lucrative dyeing industry came to an end. By around 650 CE, following the Arab conquest of the Levant, shellfish dyeing, for all intents and purposes, disappeared. Sporadic remnants lingered here and there until the fall of Constantinople in 1453, when the last traces of the dyeing process vanished. The secrets of the highly developed art, its significance once immeasurable, were lost.

The secrets remained hidden from the world for more than a millennium – until a chance event off the coast of Spain. A young zoologist stood on his boat in the Mediterranean Sea and looked with great curiosity at what a fisherman was doing. The man was showing off; he had broken open the shell of an ordinary-looking snail and was smearing his shirt with its slimy contents. Yellowish stains streaked the shirt. But as the zoologist watched, the yellow streaks slowly changed color, turning first to green, then gradually, miraculously, becoming a beautiful, brilliant purple – a purple not seen since antiquity.

The zoologist, Frenchman Henri de Lacaze-Duthiers, set sail from Minorca in Spain's Balearic Islands in 1858 to study marine life in the Mediterranean, and it is he who is credited with the rediscovery of shellfish dyeing[1]. Having seen the fisherman's captivating performance with the yellow-to-purple streaks, Lacaze-Duthiers remembered descriptions of shellfish dyeing in the works of Aristotle and Pliny the Elder. He understood that the

Henri de Lacaze-Duthiers (1821-1901)

shell the sailor had cracked open must be the same type as those once used for dyeing. He devoted years of research and experimentation to the study of these snails as well as the dyes they produced. In recognition of his contribution, he received the prestigious position of chair of the Natural History of Mollusks, Worms, and Zoophytes at France's National Museum of Natural History, eventually becoming a professor at the University of Paris.

As Lacaze-Duthiers discovered, three snail species were used for dyeing in ancient times – *Murex trunculus, Murex brandaris,* and *Thais haemastoma* – and all three belong to the same family of mollusk.

Found along the Mediterranean coast, these small creatures house an even smaller gland, part of the snail's digestive system. It is a secretion from this tiny gland that becomes the dye. From these mollusks a broad spectrum of colors can be produced, all subsumed in ancient times under the term *purple*.

Purple to the ancients meant the whole range of colors from blue to red, though today we use that term to describe only a small subset of those two hues. "Purple is a fluid chromatic concept in antiquity," writes Philip Ball in *Bright Earth,* "and the ancient dye ranged in color from bluish to a deep red, depending on how it was prepared and fixed in the cloth."[2] On the red-purple side of the spectrum falls the famous Tyrian purple, known in the ancient Near East as *argamman* while on the blue side we have the color described as biblical blue, or *tekhelet* (pronounced te-KHE-let; the kh represents the guttural Hebrew letter not found in English but corresponding to the Scottish ch as in loch or the German ch as in Bach).[3]

Each murex provides only a few drops of the precious secretion; it takes more than twenty thousand snails to produce just one kilogram of dyed wool.[4] The procedure for extracting the glands is time consuming, and the methods of dyeing are exceedingly complicated and somewhat unpredictable. It's therefore no surprise that in the ancient world the price that these dazzling purple and blue dyes could command was equally dazzling. Records show that at one time shellfish-dyed wool was worth more than twenty times its weight in gold[5]. Naturally these colors soon became a status symbol, a sign of wealth and prestige, and they embellished the robes of emperors, kings, princes, and the nobility.

THE RAREST BLUE

According to the biblical account, after the Hebrew prophet and royal advisor to the Babylonian court Daniel interpreted the mysterious writing on the wall, King Belshazzar made him the third-highest official in the kingdom, presented him with a golden chain, and robed him in purple. When Alexander the Great conquered the Persian capital, Susa, in the year 331 BCE, he found among the treasures there a number of purple robes stored for nearly two hundred years and still retaining their original luster. He admired the striking color, and soon he and his generals began wearing purple cloaks. Mordecai the Jew, who outsmarted the wicked Haman in the book of Esther, left the king of Persia's court in Susa, himself dressed as a king, adorned "in royal robes of blue" – *tekhelet* – "and white."[6] In the Gospel of Mark, prior to the crucifixion, Jesus was mockingly called a king of the Jews, clothed in purple, and crowned with a ring of thorns.

When Shakespeare, in *Antony and Cleopatra*, wanted to convey the splendor and luxury of the East, he used some of his most gorgeous language to describe Cleopatra sailing on the river Cydnus in an extravagant ship with a deck of beaten gold and oars of silver. He completed the magnificent effect: "Purple the sails, and so perfumed that the winds were lovesick with them." In ancient Rome, the use of the color purple was severely restricted; senators and other officials of the state could wear togas with purple stripes, but only the emperor could wear an entirely purple toga. The phrase "to assume the purple" described a man becoming emperor.

Every culture that came into contact with the murex dyes instantly fell under their spell, recognizing their beauty, value, prestige – and even venerability as among the Byzantine emperors. To the ancient Israelites, however, these dyes possessed a holiness not by imperial fiat but because God Himself commanded their use in His worship. Prescribed as decorative colors for the holy Temple, both purple and blue formed an important component of the ritual garb of the priests. But these colors were specified not only for public ceremonial use by the priesthood in the sanctuary. The Bible, in the book of Numbers, requires all Jews to tie a single thread of *tekhelet* to the corners of their garments as a reminder of the daily obligation to fulfill the commandments. In the Roman world, the use of distinguishing colors became increasingly exclusive, reserved for the elite, whereas in Jewish culture, the *tekhelet* string bound people together, an expression of social equality. The sky-blue color also had powerful religious significance. It evoked the vast, deep oceans, the boundless heavens, and, by association, the one infinite, unfathomable God of the Universe.

Archaeologists haven't discovered an ancient garment with the *tekhelet* thread attached to its corners – such delicate adornments not surviving the centuries – but many written sources prove the wide observance of this commandment in ancient Israel. For instance, the Talmud, the authoritative compendium of Jewish law and custom compiled around 500 CE, offers clear evidence that the Jewish people had mastered the intricacies of snail fishing and dyeing. But by the 600s, as the industry declined, murex blue became increasingly unavailable, and the Jewish practice of wearing the blue thread dwindled, eventually ceasing altogether. With the disappearance of *tekhelet,* Jews continued to tie only white fringes on their garments instead, and by the middle of the eighth century the compiler of a homiletic work known as

Midrash Tanhuma lamented, "And now we have only white, for *tekhelet* has been hidden."[7] By the time Lacaze-Duthiers made his important observations in 1858, *tekhelet* and *argamman* dyeing had long since vanished from Jewish life, their sources and techniques forgotten.

Lacaze-Duthiers's discovery, fascinating as it was, didn't lead to any practical results, however; its significance lay more in a scholarly realm. No large-scale revival of ancient dyeing methods, appropriately modernized and improved, took place. In fact, in a twist of historical irony, every form of natural dyeing – whether based on the fauna of mollusks and insects or the flora of indigo, woad, and madder – became impractical and superfluous as a result of the work of a precocious chemist whose first experiments had taken place just one year before Lacaze-Duthiers's revelation.

London's Royal College of Chemistry accepted William Henry Perkin as a student at the age of fifteen. Three years later, in 1856, he was conducting experiments assigned by his professor, August Wilhelm von Hofmann, who had theorized that it should be possible to synthesize quinine artificially. Working in an unsophisticated laboratory that he had set up in his home, Perkin tried different ways of using coal tar derivatives to produce the desired result.

Quinine, produced from the bark of the cinchona tree, originally native to the Andes mountains in Peru, at the time provided the only effective treatment for malaria. The dreaded disease, which still affects millions of people and causes almost a million deaths annually, ravaged Africa as European powers were beginning to explore and exploit the continent. The prevalence of malaria turned Africa into the "white man's graveyard," so there was every reason to research the possibility of synthesizing quinine.

Perkin didn't manage to create artificial quinine, though. It was only during World War II that two American chemists, Bob Woodward and William Doering finally succeeded. But Perkin's experiments did produce something else: a dark brown mixture that yielded the first aniline dye. We know that dye today as mauve. Not only was the hue lovely to behold, but the dye was also colorfast, which made it highly desirable. It was even called Tyrian purple when it first appeared. Perkin made a fortune from his discovery, and before long an explosion of synthetic dyes displaced the traditional natural dyes.

Interest in the forgotten art of shellfish dyeing, however, did not completely disappear. It lived on in the passionate heart of one man who dreamed of rediscovering and reviving the secret of the ancient methods. Gershon Henokh Leiner, a rabbi in the obscure Polish town of Radzyn, about one hundred miles southeast of Warsaw, served as the charismatic leader of an important Hasidic sect. But in many ways Leiner was not a typical rabbi. A brilliant scholar with a command of the vast corpus of Jewish religious texts and scholarship, he also dabbled in engineering, learned a number of languages, and taught himself enough chemistry to qualify as a pharmacologist. During the course of his studies, he developed an obsession.

Every morning, as part of the prayer service, he recited the passage from the book of Numbers that commands the Jews to attach fringes to the corners of their garments. Like Jews the world over, Rabbi Leiner thrice kissed the tzitzit – the fringes on the prayer shawl worn during the service – for each of the three times that the word appeared in the paragraph, but each kiss produced a pang. It pained him that the strings were white, while the words he fervently pronounced ordained that one of the threads be blue – *tekhelet*. How was it that for over a millennium Jews had not properly fulfilled this essential biblical law?

Dyeing strings blue, of course, didn't pose a problem. The problem was that, according to the tradition recorded in the Talmud, only dye obtained from a particular sea creature could qualify as the special blue required by scripture, and the nature of this creature, as well as the process by which its dye was produced, had been lost to history for over a thousand years. Was it possible to rediscover the lost knowledge and restore the ritual practice?

Guided by profound faith – and a willful ignorance of possible obstacles – Rabbi Leiner came to a momentous decision. He would dedicate his life to finding the authentic source of the ancient dye and to making genuine biblical tzitzit available to his people. Armed with an encyclopedic knowledge of the sometimes vague, confusing, and contradictory information in the Talmud concerning the sea creature, and familiar with much of the relevant secular, historical, and scientific knowledge, Rabbi Leiner left his disciples one day in 1887 and departed on his quixotic expedition.

His destination was the newly built Stazione Zoologica in Naples, Italy, a research center for the study of marine biology housing the Aquario di Napoli, the first aquarium ever opened to the general public. Here, if anywhere, he would be able to find the mysterious animal that produced the ancient biblical blue. Presumably he didn't know of Lacaze-Duthiers's studies of the murex, but even if he had been familiar with them, he would have rejected those snails as the genuine source. The dyes that chemists derived from the murex in the wake of Lacaze-Duthiers's findings shaded into the purple to blue-violet range, whereas Jewish tradition firmly maintained that *tekhelet* had to be a sky-blue color, evoking the heavens.

Eventually Rabbi Leiner found what he believed to be the authentic marine source of the long lost blue dye – not a shellfish at all, but a curious little ink-squirting cephalopod. Returning to Radzyn, he set up a factory to mass-produce blue strings, and within a year more than ten thousand followers proudly were wearing his cuttlefish-derived *tekhelet* on their prayer shawls.

But in another bit of irony – in a story full of ironies – early in the twentieth century another rabbi, also a brilliant Talmudic scholar and polymath, earned his PhD with a dissertation on what he called Hebrew porphyrology, or the study of purple and blue in the Bible. Rabbi Isaac Herzog's thesis proved conclusively that Leiner's cuttlefish couldn't possibly have served as the authentic source of the biblical dye. The cuttlefish's ink played no essential part in the dye process. In fact, it turned out that Leiner's dye wasn't even organic but consisted instead of synthetic material.

How could the great Hasidic rabbi have been misled? Had some unscrupulous chemist, whose help he had sought, duped him? What of the murex shells themselves? Did they conform to the traditional depiction of the miraculous creatures that once produced the sacred blue? Was it possible to reconcile the rabbinic conviction that *tekhelet* was the color of the sky with the consistently purplish hue of the murex dye?

Religion presented by no means the sole reason for modern investigation into the world of shellfish dyeing. The little sea creatures and the stunning colors they produced captivated a wide variety of individuals with quite different interests and with diverse agendas: historians and archaeologists of course, as well as biologists and chemists, but also scholars specializing in magic and superstition, numismatists, devotees of antiquarian craftsmanship, and enthusiasts who weave their own cloth and color it with murex dyes.

In the century and a half since Lacaze-Duthiers and Leiner, researchers have continued to explore the topic from every possible angle, raising a spectrum of questions. Why does an ancient Tyrian coin depict a scene featuring a dog and a murex shell? What purpose did stale urine play in the dyeing process, and why would the British Ministry of Defense fund a project to investigate murex dyeing? How do Mexican villagers "milk" a snail, and might doing so provide a solution to the ecological challenges involved in the wholesale use of murex? Could the snails' cannibal nature and dubious sexuality have any bearing on their ability to produce such exceptional dyes?

Research, experimentation, hard work, and no small measure of luck have provided many answers, though other questions remain unsolved. But in order to unravel the mysteries of these intriguing dyes, we must start at the very beginning – when shellfish dyeing was first discovered.

II
OUT OF THE BLUE

$\mathcal{N}$owadays, the most convenient way to get to Heraklion, the capital of Crete, is by air. The alternative is to go by sea; from Athens you can take a nine-hour overnight ferry or a more expensive six-hour high-speed ship. We arrived by boat in a mere two hours, having set out from the nearby island of Santorini, the remains of a massive volcano named Thera that erupted around 1600 BCE. We were on an expedition to experience the remnants of the civilization that first discovered the extraordinary art of shellfish dyeing.

A few miles south of Heraklion lie the ruins of the Palace of Knossos, an imposing architectural complex built nearly four thousand years ago where, according to Greek mythology, the great king Minos ruled. Arthur Evans, the British archaeologist who excavated Knossos in the beginning of the twentieth century, coined the term Minoan, after the legendary king, to describe the ancient civilization.

Greek myth relates that the god Poseidon gave Minos a snow-white bull as a mark of favor, and Minos, in gratitude, committed himself to offering that bull as a sacrifice. But when Minos selfishly changed his mind and decided to keep the bull, substituting another in its stead, Poseidon punished him by inspiring his wife, Pasiphaë, with a passionate love for the white bull. The union of Pasiphaë and the bull produced a monstrous creature, part man and part bull. The name of the Minotaur suggests this unnatural origin. The creature fed on human flesh, and Minos commissioned the Labyrinth, constructed by the architect and inventor Daedalus, in order to imprison the monster. Some have suggested that the palace itself was the Labyrinth, a complicated aggregation of rooms and passageways.

Even in its ruined state, the enormous site is impressive, standing as mute testimony to the advanced technological achievements of the Minoan people. It is not, however, actually a palace, but rather an administrative, religious, and professional center comprising some 1,300 rooms around a central square and covering an area of about 24,000 square meters. Air shafts allow for added ventilation, and multilevel cascading staircases maximize natural lighting. Aqueducts fed water from local springs through clay pipes to fountains, and elaborate drainage systems channeled heavy rainwater descending from surrounding hills to prevent flooding. A closed drainage system for sewage included the earliest example of a flush toilet: a seat above a drain flushed by hand with water from a jar. Beautiful colored frescos adorned the walls, and giant storage jars – still extant – held wine, grain, and olive oil. In addition to royal quarters, shrines, and storerooms, the palace contained many workshops where various craftsmen practiced and refined their expert skills.

In Phaistos, another palace not far from Knossos, the famous Phaistos Disc came to light, further proof of the highly-developed culture of the Minoans. This clay disc, fifteen centi-meters in diameter, features symbols in a spiral pattern carved on both sides. Attempts to decipher the symbols have failed so far and don't appear likely to succeed because the text sample is so limited.* The forty-five unique symbols create 241 shapes, each symbol replicated identically throughout. Since these uniform

WIKIPEDIA COMMONS

Minoan disc of Phaistos, second millennium BCE

---

* Subsequent to the first printing of this book, Dr. Gareth Owens and Prof. John Coleman claimed to have broken the code of the Phaistos Disc. They concluded that the inscription contains a prayer to a Minoan goddess. Dr. Owens' describes his discovery in a TED talk at: www.youtube.com/watch?v=6Chcplx3tZ8.
Controversy surrounding their claim has arisen, however, with some scholars rejecting their interpretation. Both sides are explored in an article in *Biblical Archeology Review* that can be read at:
www.biblicalarchaeology.org/daily/archaeology-today/phaistos-disk-deciphered/.

symbols couldn't have been carved individually by hand, scholars assume that they were made by seals pressed into the clay, which was then fired. The disc proves that the Minoans had discovered a technology for printing, or imprinting in this case, by means of movable type, beating Gutenberg to the punch by more than three thousand years.[8]

Minoan culture reached its apex around the turn of the seventeenth century BCE. Excavations at the Akrotiri site on Santorini – where we began our journey to Crete – reveal remnants of Minoan civilization in the form of inscriptions in the still undeciphered script of Linear A as well as artifacts and frescoes similar to those found at Knossos. It is possible that the massive eruption of the ancient volcano at Thera and the ensuing tsunami, as well as the arrival of the Mycenaean Greek invaders, forced Minoan civilization into decline sometime around the thirteenth century.

But given the creativity, ingenuity, and technological capabilities of the Minoans, it should come as no surprise that they were the earliest to discover the highly complex and labor intensive process of dyeing wool using secretions from sea snails. The striking blue and purple wool first produced by the Minoans became arguably the most important article of trade in the ancient world.

Both archaeological and epigraphical evidence suggest that as early as 1750 BCE the Minoans were manufacturing sea-snail dye on Crete. In recent excavations, archaeologist Robert Stieglitz found large mounds of whole and crushed murex shells all along the coast of the Aegean as well as in Knossos itself. The assumption, when a small amount of shells is found, is that the shells are related not to dye but to diet, since the snails are edible and even today form part of the menu in many coastal villages along the Mediterranean. When the amount of discovered shells is more substantial, however, as it is in Crete, the most credible explanation is that they were used for dyeing.

Furthermore, in Knossos, on a tablet that deals with textile allocations, the words *po-pu-re-jo* and *wa-na-ka-te-ro* appear. The second term translates as "royal," and the first has to do with purple, though whether it refers to cloth or dye workers remains unclear. It is the earliest written record of the shellfish dyes and attests to the prominent status that they already had attained.

Minoan art also reflects the prevalence of purple dyes. Steiglitz describes a Minoan priestess figurine from 1600 BCE, clothed in a garment with purple decorations, and a sarcophagus from around 1450 BCE that depicts men and women dressed in purple striped clothes. All available evidence indicates that *tekhelet* and *argamman* dyeing originated in the Middle Bronze Age with the resourceful and innovative Minoans on Crete.[9]

Although the Minoans are recognized as being the first to use shellfish for dyeing, a rather unlikely source has challenged that priority. Some assert that an island off the coast of Qatar was in fact the site of the earliest production of shellfish dye. The city of Al Khor on the eastern shore of Qatar lies about thirty-five miles north of Doha. The name Al Khor means "the bay" – or the sea on three sides – and its bay, open on the south to the Persian Gulf, is one of the largest in Qatar. On the eastern side of the bay, blocked from the ocean by a mangrove-lined peninsula, lies the fish-shaped Al Khor Island, or Jazirat bin Ghanim.

Today a causeway joins it to the mainland. For three years, between 1980 and 1982, archaeologist Christopher Edens investigated Al Khor Island as part of the broader six-year French Mission Archéologique Française à Qatar, directed by Jacques Tixier.

At a site on the southern side (Khor Ile-Sud), Edens uncovered five structures surrounding a central hearth, where he found pottery, flints, small grinding stones, and a single copper ring. The top layers of a large midden – the archaeological term for garbage heap – consisted of the discarded remains of sheep, goats, birds, fish, crabs, some cuttlefish, and a few shells. But the lower and oldest level yielded a surprise: It was filled almost entirely with crushed shells from one species of murex, *Thais savignyi,* totaling, according to Edens's estimate, nearly three million individual snails.

Isotopic tests to date the shells yielded inconclusive results. The earliest dating would place these shells earlier than the Minoan ones, suggesting that murex dyeing indeed originated at Al Khor. On the more reliable basis of pottery finds, however, Edens concludes that the site dates from the Kassite period of the late second millennium BCE. He maintains that around the thirteenth century BCE dyeing technology came from the Mediterranean region, where it had long been practiced. It is the first and so far the only instance of shellfish dyeing in the Arabian Gulf.

The Kassites, who hailed from the mountains of northern Iran, invaded and conquered Babylonia around 1570 BCE, heralding an era of trade and prosperity for the entire Mesopotamian region. They signed treaties with Assyria to the north and with Egypt and the Hittites to the south and west. Their sphere of influence even reached the shores of the Mediterranean.[10] Presumably the Kassites learned the art of shellfish dyeing around that time from Mediterranean dyers and brought it to Al Khor Island. The purple or scarlet material dyed there perhaps created clothes for local nobility or Kassite officials, but there is no way of knowing whether it remained a local product or became a significant export. The Kassite kingdom of Babylonia fell to the Elamites in the twelfth century.

In discussing the murex in the ancient Near Eastern world, we should mention one more archaeological find, from a period that predates the Minoan shellfish dyeing industry and even the earliest suggestion for Al Khor dyeing by five hundred years. Sargon the Great – according to some accounts the illegitimate son of a priestess and a gardener – served as a cupbearer to the king of Kish, whom he killed and whose throne he usurped. He then founded the Akkadian Empire, one of the most successful dynasties in the ancient world, which stretched from what is now Iran to the Mediterranean and from what is now Turkey to the southern tip of the Arabian Peninsula. Some scholars associate Sargon with the biblical figure of Nimrod, a "mighty hunter by the grace of the Lord."[11] When Sargon died after fifty-six years of ruling his empire, his son Rimush succeeded him, spending his entire nine-year reign in constant battle, trying to keep the empire together in the wake of widespread revolts that erupted upon his father's death.

More than four thousand years later, the Louvre in Paris acquired a partly damaged murex shell bearing the Akkadian inscription RIMUSH, KING OF KISH. The discovery of this one shell, presumably a gift to the ruler, does not, of course, prove

MARIE-LAN NGUYEN | WIKIMEDIA

Murex with inscription, "Rimush, king of Kish," dating to the late third millenium BCE.

anything about dyeing at that time, nor does it indicate the existence of any large-scale industry. Furthermore, this particular shell, which came from the Persian Gulf, not the Mediterranean, belongs to the species *Chicoreus ramosus.* It is a rare, beautiful, large murex, about one foot long, a snail fit for a king – but not for a dyer.

Seashells, such as the shiny cowries, have long served as money, decoration, or jewelry, and the shell of Rimush presumably conveyed an especially significant honor for the king. It would be extravagant to claim that this royal gift in some mysterious way indicated an early awareness of the potential use of these shells, but the connection between murex and royalty here does foreshadow the later connection between snail dyes and rulers for whom the colors signaled status.[12]

The lives of archaeologists are rarely if ever as glamorous and thrilling as Steven Spielberg's Indiana Jones movies would have us believe. The most important discoveries come only after long hours of tedious, hard work and even longer hours of research and study.

Such was the experience of a team of German archaeologists excavating a palace from the ancient kingdom of Qatna. Tell el-Mishrife, as it is known today, lies about ten miles north of Homs in Syria, in the Orontes valley. The city sits on a natural plateau, and remnants of its huge walls, 50 feet high, still exist today. The majestic palace was built around 1600 BCE and was destroyed when the Hittite kingdom of Anatolia in modern Turkey reached its height and conquered Syria in 1340 BCE. Other archaeologists had excavated the site before, but in 2004 the German team discovered an underground tomb cut into the cliff beneath the main edifice. Sealed and forgotten, this room had not had any contact with the elements, leaving its contents exceptionally well preserved.

Workers retrieved over two thousand artifacts from this remarkable find, including jewelry, pottery, statues, and human and animal bones. But that wasn't all the team uncovered. In their detailed examination of the tomb, archaeologists noticed colored flecks on the crypt floor. Intrigued, they sent the soil for chemical testing, and the results showed that the patches of color contained a dye that came from murex snails. It was exciting news.

The team returned to the tomb floor, meticulously hand sifting through all the dirt, looking for scraps of fabric. Eventually they found a few thousand tiny bits of cloth, most only a fraction of an inch, and of those only a small number were colored. These scraps, perhaps the remains of ancient burial shrouds, represent the oldest specimens of shellfish-dyed fabrics found so far.

The earliest occurrence of the familiar terms for the blue- and purple-dyed wool – *tekhelet* and *argamman* – appears in connection with an ancient royal marriage. The wedding of Tadukhipa, daughter of the Mittani king Tushratta, and the Egyptian pharaoh Amenhotep III took place sometime around the fourteenth century BCE. As expected, the bride in this no doubt political union arrived in Egypt with a substantial dowry from the land of the Mittani, today northern Syria and southern Turkey. Cuneiform script on clay tablets duly recorded the details of this generous gift. Those tablets lay buried under the hill of Amarna – about halfway between Luxor and Memphis – until an old woman gathering fertilizer for her garden discovered them in 1887. There, among the various catalogs and items described in the more than 350 tablets that comprise the Tel El Amarna Letters, we find the entry – *sabattu sa takhilti* – "a dress [or sash] of blue." "*Argamannu*," or purple-dyed cloth, also appears.[13] These two colors represent the biblical blue and Tyrian purple of ancient times. Clearly by this time *tekhelet* and *argamman* had become well-known and valued items, fitting gifts for pharaohs and kings.

The ancient Israelites, when they adopted monotheism, took what the world at large considered precious and majestic, and consecrated it to the service of God. The Bible first mentions *tekhelet* and *argamman* in Exodus, chapter 25, as part of the list of materials required for building the Tabernacle, the portable temple erected and dismantled over and over again during the forty-year wanderings of the Israelites in the wilderness. The list also includes other valuable commodities such as gold, silver, copper, incense, semiprecious stones, and sealskins. The walls of this traveling temple consisted of curtains made of fine linen and of wool dyed blue, purple, and scarlet hung from long beams fashioned from wood and overlaid with gold. A design depicting heavenly cherubs decorated the curtains of the Tabernacle.

Scarlet, called *tola'at shani,* or "crimson worm," in the Bible, known today as kermes, is the red dye produced by drying out and then crushing certain insects. A drape or veil woven of the same trio of colors – *tekhelet, argamman,* and *tola'at shani* – divided the Tabernacle into two parts. On one side of the veil stood the golden objects, such as the incense altar and the candelabrum used as part of the daily service. On the other side, the Holy of Holies housed only a single item: the Ark of the Covenant. The most sacred of all vessels in the Temple, the ark contained the two tablets inscribed with the Ten Commandments that Moses had received on Mount Sinai. When the Israelites broke camp and resumed their travel in the desert, the Levites – the priestly clan among them – wrapped the ark in a covering made completely of *tekhelet* and carried it on their shoulders to their next destination.

The high priest who performed the rites in the Tabernacle, and later in the Temple in Jerusalem, wore elaborate garments also woven from these precious dyed fabrics, including a long tunic of pure sky blue over a white shirt and trousers. The same striking contrast of blue and white was mirrored in the tzitzit, the tassels or fringes made of *tekhelet* attached, along with white strings, to the corners of the hem of each layperson's clothes. The book of Numbers states:

> *The LORD said to Moses as follows: Speak to the Israelite people and instruct them to make for themselves fringes on the corners of their garments throughout the ages; let them attach a cord of blue to the fringe at each corner. That shall be your fringe; look at it and recall all the commandments of the LORD and observe them, so that you do not follow your heart and eyes in your lustful urge. Thus you shall be reminded to observe all My commandments and to be holy to your God.*[14]

The blue thread on the corner of one's clothes serves as a reminder not to go astray, or in the words the ancient Confucian proverb: "I hear and I forget, I see and I remember."[15]

Fashion designers often obsess over hemlines, meticulously calibrating each season's ever changing optimal length according to some secret formula that only they understand. In ancient times the hem had just as much significance, not because of its length, but rather, to quote the scholar Jacob Milgrom, because the hem of one's garment was an extension of his "person and authority."[16] In the Bible when David – not yet king of Israel but a young fugitive fleeing King Saul – furtively cut off the hem of Saul's cloak, the old king correctly construed the act as a threat against his own leadership: "I know now that you will become king, and that the kingship over Israel will remain in your hands," he exclaimed.[17]

In ancient Akkadia, an exorcist recited an incantation over a person's severed hem as part of the rite of "cutting off the hem" to chase away demons. A husband could divorce his wife by cutting off the hem of her robe. The Mari prophets of Mesopotamia sealed their letters to the king with a lock of hair and a piece of their hem, thus binding themselves to the veracity of the contents. When a decorative tassel, which often dangled from the corner of a person's clothes, was pressed into wet clay, the resulting impression served as a signature on legal documents.

To this day, a person called up to read from the Torah in a synagogue touches his tzitzit to the scroll and then to his lips, affirming his commitment to the Torah's teachings. That the tassel and hem reflect their owner's personality also appears in the custom among some Jews of Holland to hand embroider a personal design on the corner of their prayer shawl that will be worn for the first time on their wedding day.

Tekhelet is the color of the Temple; its curtains, fabrics, and decorative coverings were predominantly blue, as were the clothes of the priests who served there. But rather than excluding the layperson from this elite society, the Bible calls upon the entire community to join that religious aristocracy. By attaching a bit of the sacred, the *tekhelet* thread, to his everyday clothes, each individual becomes, in effect, a priest. "You shall be to Me a kingdom of priests and a holy nation," the Bible states.[18] The tzitzit illustrate, in the words of Milgrom, "the epitome of the democratic thrust within Judaism which equalizes not by leveling, but by elevating: all of Israel is enjoined to become a nation of priests."[19]

Tekhelet appears throughout the Bible, and, as with Mordecai in the Persian court and Daniel before the Babylonian king, it always indicates royal status, prestige, and wealth. Ezekiel, for example, describes the imposing Assyrians, "clothed in blue, governors and prefects, horsemen mounted on steeds."[20]

The book of Judges recounts how Deborah the prophetess led the Israelites to victory in battle against the oppressive Canaanites. In one particularly moving scene, the mother of Sisera, the enemy general, sits at the window with her ladies in waiting, sobbing and wondering what is keeping her son so long at the battle. She doesn't know yet that he lies dead, his army demolished at the hands of the Israelites. Instead, she convinces herself that he has not yet returned because he is busily dividing up the victory spoils, among them probably the dyed fabrics that made the region famous: "Spoil of dyed cloths for Sisera, of embroidered cloths, a couple of embroidered cloths round every neck as spoil."[21]

The reasons for going to war, according to this passage from the book of Judges, included looting the precious dyed fabrics that the enemy manufactured. Perhaps an unspoken goal of Sisera's military campaign was to gain control over that important and profitable local dyeing industry. The biblical text stresses that specifically the northern tribes, Naphtali and Zebulun, fought in this battle. Before the Israelites had entered the land, Moses had blessed all the tribes, and his farewell to Zebulun concluded with the passage, "For they draw from the riches of the sea and the hidden hoards of the sand."[22] The later rabbinic interpretation of that phrase claims that the hoards hidden in the sand refer to the *hillazon,* the term used by the Talmud to describe the marine source of the blue and purple dyes. Ultimately deriving from the

Greek word *helix* ('ἑλιξ), which means spiral, *hillazon* specifically denotes a snail with its spiraling shell. According to the Talmud, the marine *hillazon* is the only legitimate source of the *tekhelet* dye.

During this period, 1550–1200 BCE, the Late Bronze Age, episodes involving the blue and purple dyes crop up all over the Near East. An event recorded in a large stone tablet known as the Manapa-Tarhunta letter, discovered in the 1980s in a pile of trash in an ancient temple, reveals a fascinating and mysterious tale of political intrigue.

In Anatolia, one of the ancient world's most notorious characters, Piyamaradu, reigned as king of the coastal region of Wilusa. Scholarly conjecture identifies him as someone the world knows far better: King Priam of Troy. Piyamaradu's neighbors to the east, Manapa-Tarhunta, king of the Seha River Land, and Muwatalli II, ruler of the Hittites, each sent a group of "SARIPUTU-men" to the island of Lazpa (today's Lesbos), then under Trojan rule. Piyamaradu, unprovoked, abducted the two teams. Evidently, the apple didn't fall far from the tree; according to myth, Priam's son Paris later enacted the most famous abduction of all time, the kidnapping of Menelaus's beautiful wife, Helen, which of course sparked the Trojan War.

But who were these sariputu-men, and what were they doing at Lesbos? The words of the hapless sariputu-men appear in that letter: "We are tributaries, and we came over the sea. Let us render our tribute," the emissaries beseeched Piyamaradu. The Hittite word for "tribute" here is *arkamman*. Professor Itamar Singer of Tel Aviv University believes that this word is in fact the same as the Akkadian and Hebrew *argamman,* namely purple wool, and therefore proposes that the sariputu-men were purple-dyers. Perhaps they had come to Lesbos to practice their art. If this is the case, we have a picture of traveling purple-dyers – in a time before the industry had established permanent dye facilities – who journeyed from site to site to collect the mollusks and dye on the spot. In the ancient world it was not uncommon for craftsmen to be brought from one palace where they were employed to another.

It is also possible that they went to the island, home to a renowned deity, on a politico-religious mission: to drape the goddess of Lazpa with purple robes in a form of adoration practiced at that time. The prophet Jeremiah, for example, mocks idolaters who worship the work of their own hand and clothe their idols in blue and purple.[23] A poetic fragment by Sappho, the Greek lyric poet of Lesbos, mentions a purple headdress presented to the goddess Aphrodite.[24] But whatever the motivation, these purple-dyers, important men, came under the protection of their kings, which made Piyamaradu's act all the more insolent.

Professor Singer deduces from the tone and content of the letter that "representatives of the Hittite crown closely supervised the movements of these itinerant craftsmen and controlled their lucrative revenues."[25] Intervention and diplomacy won the day, and Piyamaradu released the dyers, but tensions between the parties survived in war and counterwar – until the Hittites eventually sacked Troy.

In the Holy Land, warfare between the Canaanites and Israelites raged hard and long until a rapprochement allowed the two to live side by side in relative peace in the land of Canaan. Though the origin of the name Canaan (*kn'n*) remains unclear, one theory again points us toward shellfish dyes. Some scholars propose that the word actually meant "purple," so that the term applied originally to the dye, then later to the people, but it's equally possible that the term attached first to the people and later described the dye that they produced.[26] Avid traders, the Canaanites dealt in all kinds of wares and goods, from spices to dyed fabrics; eventually the term *Canaanite* itself came to mean "merchant" or "trader."

The book of Proverbs depicts an imagined Woman of Valor, a superwoman of sorts, who among her many skills "makes cloth and sells it, and offers a girdle to the merchant," where the word translated as "merchant" is *Canaanite*.[27] The prophet Zechariah speaks of the Day of the Lord and the idyllic Messianic era that will ensue when all people will come to sacrifice to God in the Temple. On that day, "there shall be no more traders in the House of the LORD of Hosts," where Caananite, translated as trader, refers to money changers in the Temple.[28]

In the earliest biblical times, the Canaanites lived primarily in the lowlands, and the Israelites inhabited the mountains. When the Sea Peoples – presumably the Philistines – invaded the region, they drove the Canaanites north into what is today Lebanon and Syria. At the time of King David (around 1000 BCE), they had become the leading force in Mediterranean culture and commerce. By that time, the Canaanites living along the eastern coast of the Mediterranean were becoming known by a new appellation: "Phoenician."[29]

III
PURPLE PEOPLE

$\mathscr{S}$ometime during the second century CE – or possibly earlier or later – Dionysius Periegetes, a Greek author who may or may not have come from Alexandria, wrote a long poem in the style of the Homeric epics. That is the one indisputable fact about him. The poem, *A Description of the Inhabited World,* mixes geographical fact and fancy, and remained popular for centuries. This is what he writes about the Phoenicians:

> *Upon the Tsurian sea the people live*
> *Who style themselves Phoenicians...*
> *These were the first great founders of the world –*
> *Founders of cities and of mighty states –*
> *Who showed a path through seas before unknown.*
> *In the first ages, when the sons of men*
> *Knew not which way to turn them, they assigned*
> *To each his first department; they bestowed*
> *Of land a portion and of sea a lot,*
> *And sent each wandering tribe far off to share*
> *A different soil and climate. Hence arose*
> *The great diversity, so plainly seen,*
> *'Mid nations widely severed.*

Dionysius got a few things right. An ancient people, the Phoenicians did indeed found city states – though others had done so before them. These enterprising and daring seafarers certainly established colonies far from their native soil. Renowned maritime traders who did business with anyone and everyone with whom they came in contact, including the Greeks and the Egyptians, the Phoenicians traded everything from wine and glass to hunting dogs and salted fish.[30]

They were also the first to make extensive use of an alphabet, and their alphabet is the ancestor of most modern alphabets, including the Arabic, Hebrew, Greek, Latin, and possibly even Indian. Ancient scripts that existed prior to the Phoenician alphabet made writing difficult to learn and inaccessible to the masses. Egyptian hieroglyphs, for example, consist of a system of highly complex pictographs that represent concepts or sounds. Writing remained the exclusive domain of an elite class of scribes, frequently priests, who used their specialized knowledge to help dominate society.

The Phoenician alphabet, on the other hand, revolutionized the written word by employing only one symbol to represent one sound – though those symbols lacked representations of vowels, which the Greeks later supplied as their contribution. Easy to learn, the alphabet spread quickly along the far-flung trade routes developed by the Phoenicians and greatly simplified the act of bookkeeping. Much ancient writing, whether cuneiform or hieroglyphic, consists of records – of ownership, supplies, or taxes. Simpler styles of writing made for an obvious advantage in that respect, especially for a commercial society. Along the trade routes they established, the Phoenicians presumably also encountered the dye sites of the Minoans. Always enterprising, they soon learned the dyeing techniques and in time took over the industry, setting up dye houses all along the Mediterranean shores and eventually establishing blue and purple dyed fabrics as their leading articles of trade.

Tyrian coin depicting Melkarth's dog discovering the murex, 3rd Century CE

COURTESY OF THE KADMAN NUMISMATIC PAVILION ERETZ-ISRAEL MUSEUM, TEL AVIV

The legendary origin of Phoenician supremacy in the art of dyeing, though, offers a far more romantic story. The Greeks and Romans, who later retold the tale, identified the Phoenician god Melkarth, whose name means "king of the city," originally god of the city of Tyre, as Herakles or Hercules. Melkarth was strolling on the beach one day with one of his many lovers, the attractive nymph Tyros, and trotting along ahead of them went their sheepdog. When the dog innocently bit into a shellfish and looked up, a deep, bright purple had stained his muzzle. Tyros begged Melkarth to make her a gown dyed in this beautiful hue that no one had ever seen before. Eager to keep her favors, he complied and immediately began gathering shellfish in order to produce the newly discovered dye. Tyre, one of the major cities of Phoenicia, supposedly derives from the name of the nymph Tyros – though the name actually means "rock" – and Tyrian purple became a much desired fabric in the ancient world.[31]

The Phoenicians established colonies all over the Mediterranean to mine tin and other natural resources and also to collect murex and manufacture dyed wool. These colonies grew into ports of trade, and their inhabitants traveled as far as Sicily, Malta, Tunisia, Spain, and – according to the Greek geographer Strabo – even Britain. Some extreme enthusiasts claim that the Phoenicians were the first to sail to America or even that they reached as far as Australia. Less controversial is the fact that some of the Mediterranean's major ports began as Phoenician commercial outposts: Tripoli, Genoa, Palermo, Algiers, Ibiza, Cartagena, Málaga, Gibraltar, Cádiz, and Tangier.

The largest of the Phoenician colonies, though, was Carthage in North Africa, in what is now a suburb of Tunis. Founded in the ninth century BCE, Carthage's star rose until it challenged the Roman Republic for dominance of the ancient Mediterranean world. In the second century BCE, the Punic Wars settled the question of regional supremacy once and for all – "Punic" being the English version of the Latin word for "Phoenician" – when Rome prevailed and Carthage fell. The Roman victors passed laws prohibiting the rebuilding of the city, but in the days of Julius Caesar, when the Roman Empire was taking shape, Rome itself built a new city on the site, which became a major commercial center.

Symbol of Tanit
JUAN ANTONIO RUIZ RIVAS | WIKIMEDIA COMMONS

The history of Carthage also gives us a glimpse into a darker, lesser-known side of Phoenician culture and religion. The Greek historian Diodorus Siculus recounts that as part of the worship of the goddess Tanit there stood in the city of Carthage "a bronze image of Cronus

extending its hands, palms up and sloping toward the ground."
Upon this statue a priest placed a young child. "Each of the children when placed thereon, rolled down and fell into a sort of gaping pit filled with fire."[32] Child sacrifice seems to have taken place in other Phoenician cities as well, though some scholars deny this and reject the traditional interpretation of the so-called Sign of Tanit: a drawing, found in many Phoenician cities, of a triangular stick figure with a circular head and hands upturned that may refer to this particular form of human sacrifice.

What Dionysius Periegetes got wrong was the notion that the Phoenicians styled themselves "Phoenicians." No nation called Phoenicia ever existed; much like the nearby Hellenic world, Canaanite-Phoenician civilization formed around independent city states, often allied with each other but sometimes in conflict. People regarded themselves as Tyrians or Sidonians rather than members of a broader national group, and the Tyrians who founded Carthage gave it that name – *Qart-Hadašt,* which means "new city" – because they were founding what was, to them, a new city of Tyre. Indeed, the word *Phoenicia* derived from the ancient Greeks in reference to the region of the shellfish dyeing industry and the purple and blue wool universally associated with the Phoenicians.

The etymology of the word *Phoenicia* has given rise to much discussion, not always absolutely definitive. The accepted history of the word is that it comes – through *phoînix* (φοῖνιξ), meaning Tyrian purple – from *phoinos* (φοινός), meaning blood red. Some linguists propose that the Greek word *phoînix* actually has its origin in the Semitic languages spoken by the Canaanites and Israelites who mastered the dye process.

The Hebrew word *puwwa* and Ugaritic *pwt* mean red-madder dye. Issachar, one of the children of Jacob in the Bible, named his sons Tola and Phuvah, which has led to the suggestion that these clans were dyers – *tola* suggesting *tola'at shani,* or crimson worm, and *puwwa,* another term for dye. It has further been suggested that *puwwa* relates to the Hebrew *parper* and Arabic *farfara,* which mean to boil or bubble. The onomatopoeic overtones in the word *purple,* as in the word *bubble,* evoke the sound of a boiling cauldron. Following this line, the etymology would have evolved from the word for boiling, a part of the dye process, to the application of that word specifically to dyeing and eventually to the murex-based red-purple and blue-purple dyes themselves.

Whatever the exact origin of their name, the Phoenicians formed close political alliances with their neighbors to the south, the Israelite kingdom. Hiram, the Phoenician king of Tyre, sent King David materials as well as workmen to help him build his palace. Later, David's son Solomon looked to Hiram for help when he undertook his crowning achievement, the Temple in Jerusalem. Solomon collected the vast resources required, searching far and wide for the finest raw materials and the most skillful craftsmen to help with this tremendous task. King Hiram sent his famous cedars of Lebanon to Jerusalem, along with the expert dyers and weavers of *tekhelet.*[33]

The alliance between Tyre and Israel proved strong and enduring. In fact, the Israelite king Ahab even married a Phoenician princess, Jezebel, and they ended up being the most notoriously wicked royal couple ever to rule Israel, zealously advancing the worship of Baal, the Tyrian god. No doubt dressed in her native Tyrian purple, Jezebel became the prototype of the promiscuous, seductive, destructive woman, a danger to every man around, a biblical femme fatale.

Hundreds of years later, the prophet Ezekiel recounted the glory of the Phoenicians as the greatest maritime traders who dominated the fabric industry: "Of blue and purple from the coasts of Elishah were your awnings."[34] But he took them to task for their hubris and greed: "You grew haughty because of your beauty, you debased your wisdom for the sake of your splendor," predicting in brutal detail their ultimate downfall.[35] The threat to the Phoenicians and their Hebrew neighbors, though, arose from unexpected quarters. For centuries the balance of power in the Near East shifted back and forth between the Assyrians to the north and the Egyptians to the south.

But the winds were changing, and a storm was brewing from the east.

The Babylonian king Nabopolassar revolted against the Assyrians in 612 BCE and destroyed the Assyrian capital city of Nineveh. A few months before his death, he sent his son Nebuchadnezzar to fight one of the most significant battles in ancient times, the Battle of Carchemish. The Babylonians, upstart claimants, were challenging the balance of regional power. On the west bank of the Euphrates, at what is now the border between Syria and Turkey, in the summer of 605 BCE the great powers gathered for a war that determined the course of events for years to come. The Assyrian king Ashur-uballit II had formed a political alliance with the king of Egypt, Pharaoh Necho II, to combat the rising Babylonian threat. Josiah, king of Israel, allied himself with the Babylonians – in what turned out to be an ill-fated partnership.

In order to prevent or at least delay Pharaoh Necho's armies from reaching Carchemish, Josiah engaged the Egyptian king in battle at Megiddo. The devastation that ensued, including Josiah's own death, gives us the word *armageddon* – from the Hebrew *Har Megiddo,* the mount of Megiddo. Devastating as it was, though, the delaying tactic fulfilled its goal, and Necho arrived in Carchemish too late to join the battle. Nebuchadnezzar had already captured the city, putting a permanent end to the Assyrian empire. Free now to focus on the Egyptians, Nebuchadnezzar thoroughly defeated them, too, relegating Egypt to the status of second-rate power. Babylon now ruled the Near East.

In alliance with the Phoenicians of Tyre and other neighbors, Josiah's successors on the throne of Judah foolishly rebelled against their former Babylonian allies. In 597 BCE, Nebuchadnezzar forcibly exiled Judah's king along with the aristocracy, and in 586 he destroyed the city of Jerusalem, burned the Temple, and tortured, murdered, or exiled much of the surviving population of Judea. The wealth and treasures of Jerusalem and of the Phoenician cities that also fell to Nebuchadnezzar made their way back to the coffers of the Babylonian king. "All the vessels of the House of God, large and small, and the treasures of the House of the LORD and the treasures of the king and his officers were all brought to Babylon."[36] Purple and blue cloth made for the greatest prize, as cuneiform annals testify: "Colored garments, linen stuffs, blue and purple stuffs" – *kitû subâtu* **ta-kil-tu** *subâtu* **ar-ga-man-nu** – "ushu wood, ukarinu wood, everything costly from the royal treasure."[37]

The spoil of textiles dyed from the murex sea snails traveled more than a thousand miles from their origin in the Mediterranean to the Babylonian kingdom in what is today Iraq. But their journey would take them ten times farther yet, to the very edge of the Asian continent.

"Written laws are like spiders' webs and will, like them, only entangle and hold the poor and weak, while the rich and powerful will easily break through them."[38] The Scythian philosopher Anacharsis spoke these words 2,500 years ago, and great writers and thinkers from Jonathan Swift to Friedrich Nietzsche have echoed them ever since. Following Strabo, Pliny the Elder credits Anacharsis with the invention of the two-armed anchor, and legend has it that he invented the potter's wheel.[39] Influential and innovative though he may have been, he didn't exactly represent Scythian culture, philosophy not generally being a strong point of that society. In fact, when Anacharsis returned from his studies in Athens, his own brother murdered him, an act more in line with what we know of the Scythians.

The finer details of their origins come mostly from legend, but it appears that the Scythians migrated from what today is northern Iran toward Ukraine sometime around the eighth century BCE. Assyrians called them the *Iskuzai,* and the Bible refers to them as *Ashkenaz.* Allies of the Babylonians, they joined with Nabopolassar and Nebuchadnezzar in their battles against Assyria. Most of what we know about the Scythians and their culture comes from the Greek historian Herodotus, not always the most reliable of sources but always ready with a good story. He relates that they lived a nomadic life, were excellent horsemen, and are credited with taming the horse and turning it into a weapon of war. The Homeric poems call them "the horse-milkers,"[40] while Herodotus describes them as wearing tunics and padded trousers, which they tucked inside boots.

At birth, they cauterized a baby girl's right breast so that her muscles wouldn't weaken and she could better handle a sword – according to legend, anyway. The Scythians believed in equality of the sexes, and their women, heavily tattooed just like the men, participated in war. The characteristics of these fighting women amalgamated somewhat with Greek myths about the Amazons, a race of female warriors, which some mythographers placed in Scythia.

Before a Scythian woman could marry, she had to kill three enemies in battle – no doubt leading to an interesting singles scene on the steppes of central Asia. The Scythians didn't have saddles, and men and women rode only with saddlecloths. Skilled in guerilla warfare, they used barbed and poisoned arrows, and their fearsome martial skills contributed to the belief that the Huns descended from them. The words of Jeremiah the prophet are said to refer to the Scythians: "Lo, he ascends like clouds, his chariots are like a whirlwind, his horses are swifter than eagles. Woe to us, we are ruined!"[41]

Before burying him, the Scythians mummified the body of a dead king. On seeing the royal corpse, every man in the tribe had to sever a piece of his ear and drive an arrow through his left hand. After the ruler's funeral, they purified themselves by entering a yurt filled with smoldering cannabis, inhaling the smoke, a ritual that certainly did much to help them forget the pain of their wounds. But the ceremonial tribute didn't stop there. A year after his death, fifty of the king's best servants were strangled, along with fifty of his most beautiful horses. A huge circular grave was dug around the king's burial mound where all the mummified horses and riders were positioned as if still galloping.[42]

The most famous of the Scythian burial mounds, called kurgans, to be excavated in modern times were found in the Pazyryk valley. Between 1925 and 1949, the so-called father of Russian archaeology, Sergei Rudenko, uncovered five large kurgans and several smaller ones in the Altai Mountains of Siberia, south of Novosibirsk and north of the four-way border among Kazakhstan, China, Russia, and Mongolia. The Pazyryk excavations shed much light on Scythian culture, such as their skill in crafting spectacular gold works of art. But it was the condition of the find that made the Pazyryk artifacts even more valuable. The geology in that region is permafrost, where only the top layer of soil thaws in the summer. The Scythians would bury their dead, and then build mounds on top of those graves, effectively insulating the burial layer, which would now stay frozen year-round. When Rudenko uncovered the ancient relics, he found them in a nearly perfect state, having been both sealed and frozen.

The Pazyryk artifacts, now located in the State Hermitage museum in St. Petersburg, include a complete mummified tribal chief, his tattooed skin still visible. In addition to carriages, musical instruments, golden jewelry, adornments both for women and horses, mirrors, tools and utensils, and, of course, weapons, Rudenko also found fabrics, the earliest dating to the fifth century BCE, among them the oldest pile carpet in existence.

An even more significant find was a saddlecloth measuring 235 by 60 centimeters. The large textile, made of felted wool, has an intricate woven pattern which incorporates design elements indicating that it was made in the Near East. For example, in some places the usual warp and woof threading is replaced by threads aligned in only one direction held together by loops of gold, typical of Assyrian or Babylonian artistry.[43] The central design consists of evenly spaced rows with decorative squares arranged on a background of rich purple. The surrounding border contains delicate images of kings and queens adorned with crowns and royal garb on a background of sky blue. Small golden rectangles line the edges, and five purple tassels hang down from wooden pieces along the bottom. When subjected to chemical analysis, the dyed wool revealed a mixture of three chemical compounds: indigo, monobromoindigo, and dibromoindigo. The presence of brominated indigo definitively demonstrates that the dyes came from murex snails, since these snails are the only source of those components.

How did murex-dyed wool from the Mediterranean make its way a thousand miles east, woven into a tapestry in Babylonia, and from there north into the mountains of central Asia? Ancient trade routes connected Tyre and Babylon and Babylon with Scythian cities to the north. Jewish legend also has it that, after sacking the Temple and exiling the inhabitants of Israel, Nebuchadnezzar left behind the snail catchers and *tekhelet* dyers to continue their craft, the proceeds going to the Babylonian king.[44] Such an arrangement typified how conquerors in the Near East treated the production system of the purple and blue dye. The precious system formed a significant part of any war's plunder in the region, each victor bringing the highly lucrative industry under his control.

Perhaps conquering Babylonian soldiers carried home the constituent wool of this Pazyryk saddlecloth, the oldest patterned cloth of *tekhelet* and *argamman* in existence. In Babylon, skilled artisans wove it into a splendid cloth that then made its way north into the hands of the Scythian chief, whose kurgan Sergei Rudenko later unearthed. After two and a half millennia, the vivid colors remain vibrant and crisp, demonstrating the incredibly steadfast nature of the dye.[*]

---

[*] More recent research and analysis of the Pazyryk saddlecloth has called into question whether the blue wool is plant- or snail-based indigo. There is no doubt however, regarding the purple, which certainly came from murex.

IV
EXPLORING DORA

$\mathcal{H}$e reluctantly gave the order: Dump all the cannons, weapons, and extra ammunition into the sea. On the night of May 21, 1799, General Napoleon Bonaparte made the difficult choice to abandon his precious artillery after the ships with which he was to rendezvous for an evacuation failed to show. Horses and carriages previously used to transport arms went into service to evacuate the wounded. The dead and dying were everywhere, French diplomat Louis de Bourrienne recounted in his memoirs. "On our right lay the sea; on our left, and behind us, the desert made by ourselves; before were the privations and sufferings which awaited us. Such was our true situation."[45]

As Napoleon stood in the shallow waters watching his men ferry out cannons on makeshift rafts and haul them overboard, he probably took no notice of the small shells beneath his feet. But the role that those shells played on that beach had a greater historic significance than the retreating forces of the Corsican general from the French Republic. The beach covered that night in blood and gunpowder was called Tantura, but for close to three thousand years it had been known as Dora.

Napoleon had sailed to Egypt and occupied it in the context of his struggle with the British Empire for world domination. His strategy aimed to disrupt British supply routes to India. After suffering major defeats at the hands of Admiral Nelson and the British Navy, he turned his sights to the Holy Land, hoping to cut off any possible land attack on Egypt by the British, who would surely try to use the ports of Palestine and Syria to offload troops and supplies.

The campaign proved an unmitigated disaster. After Napoleon failed to take the fortress at the ancient city of Acre, he had to retreat. He had hoped to use Dora, thirty miles south, as his evacuation port, but Rear Admiral Jean-Baptiste Perrée, seeing two British ships in the vicinity, sailed directly back to Europe instead of rendezvousing with Napoleon as originally planned. Napoleon turned farther south and took the sick and wounded to the port of Jaffa, where other French ships sat at anchor.

Most of the guns, cannons, and even some ships that tell the story of that fateful night still lie underwater off the coast of Dora, known today by the Hebrew name Dor.

Some of the artifacts, however, are on display at a local museum a few yards inland called the Glass Factory. The museum building, which houses relics from the three-thousand-year history of the city, indeed served as a glass factory in the late 1800s, almost a hundred years after Napoleon's debacle – albeit briefly.

Baron Edmond de Rothschild, always seeking opportunities to increase the self-sufficiency of the Jews of Palestine, founded a winery at the nearby town of Zichron Yaakov in 1885. Hoping to manufacture the requisite bottles from the sands of the nearby beaches, he enlisted the help of a young chemical engineer studying in Paris at the time, Meir Dizengoff – later Tel Aviv's first mayor – and sent him to Dor to set up a factory to produce the glass. The sand at Dor, however, proved unsuitable for melting to glass. After several failed attempts to import sand from Italy, Rothschild cut his losses and abandoned the venture. Of Dizengoff's glassworks only the quaint building remains, with its thick stone walls and sloping red tiled roofs. It once contained batch mixing systems and furnaces used in the production process. On permanent display in the Glass Factory museum is an exhibit devoted to a much older, more successful industrial enterprise: the ancient industry of shellfish dyeing, for which in Phoenician days Dor was renowned.

Within easy view from the large wooden doors of the Glass Factory museum, the cliff of Tel Dor juts out into the ocean. There the ancient Phoenician city of Dor once flourished. The rock in the area is an unusual type of sandstone called kurkar – a soft material easily carved and quarried – formed from fossilized underwater sand dunes and unique to the northern Israeli coast. Five kurkar ridges run parallel to the coast, two completely submerged and three partially or fully exposed, including the ridge on which lies the tell itself – the mound of the accumulated remains of an ancient settlement. The kurkar ridges correspond to the various positions of the shoreline throughout geological history.[46] Over the millennia, ocean waves have sculpted sharp crags, caves, and open-ended tubes that form geysers in the area. Sea activity has also created another geological curiosity: a coastline dotted with small islands surrounded by shallow shelves. These formations resemble high hats with a wide brim – a type of Druse headdress known as a *tantura*, which is what the local villagers called the region.

To the south of the tell, a string of small islets forms a natural breakwater that frames shoals and lagoons. Along these islets, both natural and man-made pits fill with seawater – or salt where seawater has evaporated – so it's easy to see why Dor also produced another key commodity in the ancient world: salt.* In his popular microhistory *Salt,* Mark Kurlansky suggests that the Romans first took an interest in the Phoenicians' purple dye when they took over the salted fish trade. In some areas, the kurkar, swept relentlessly by sea and wind, forms razor sharp grooves. In others, the ground teems with large patches of shells that collected there under the force of some long-gone current. Between the tell and the largest islet, called Shehafit after the terns that nest there, lies a charming ultramarine lagoon, the southern harbor for the ancient city.

Situated between Haifa to the north and Jaffa to the south, Dor served as one of the most important ports in ancient Israel. Its natural lagoons and coves offered ships safe harbor and dock, and it sat on the hugely important Via Maris, the sea road trade route that connected Egypt through the Carmel ridge past Megiddo to Syria and Mesopotamia in the east. Archaeologists have been digging at Dor for almost a century, with the most extensive work carried out between 1980 and 2000 by Ephraim Stern. They have uncovered the rich history of a city inhabited progressively by Canaanites, Sea Peoples, Israelites, Phoenicians, Assyrians, Greeks, Romans, Byzantines, and Crusaders.

---

* The pits on these islands can be deceptively deep. On one excursion, my then-four-year-old son stepped into what he thought was a shallow hole and had to be scooped up before drowning.

Dor appears in the book of Joshua as a royal city of the Canaanites and later as part of a district that Solomon gave to a son-in-law, Ben-Abinadav, to govern. The city for the most part escaped the destruction that much of the region suffered during the campaigns of Assyria and Babylon in the seventh and sixth centuries BCE. As the center of the region's political gravity shifted from Babylon to Persia to Greece, Dor continued to play its role as a commercial and cultural center, although its importance as a strategic port meant that battles or sieges could not be avoided. During the Classical period, the Athenians captured Dor from Sidon, a Phoenician city allied with Persia, and it served as the farthest outpost of their navy.[47]

Though a crucial port city frequently at risk of siege, Dor did enjoy interludes of peace and prosperity. Such was its status during the Persian and into the Hellenistic period, when Dor richly combined Phoenician and Hellenic culture. Commercially, much like any other bustling port around the Mediterranean, its inhabitants engaged in shipbuilding and fishing, buying and selling. Dor stands out, however, for the prominence of one key industry, namely the manufacture and trading of shellfish-dyed wool.

Archaeologists have uncovered evidence of an extensive and elaborate production enterprise that involved the collecting and preserving of massive quantities of murex snails and the preparation of dye on an industrial scale. Below the western side of the city's cliff lie large rectangular pools quarried just below sea level that may have served as holding pens to store the murex and keep them alive until the dyers were ready for them. The ubiquity of crushed murex attests to the scale of the endeavor, and the residents of Dora even incorporated these broken shells into the

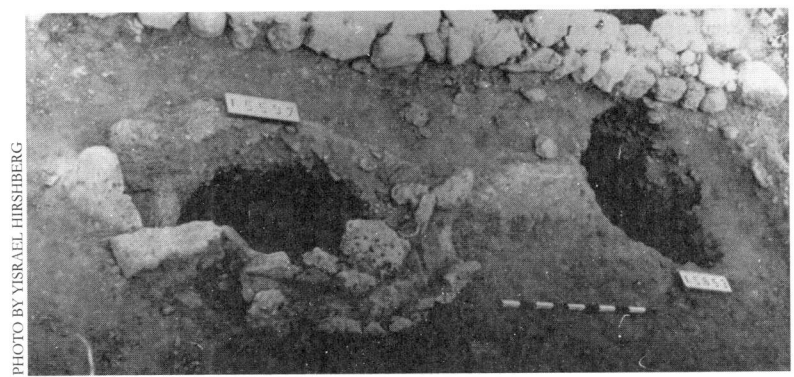

PHOTO BY YISRAEL HIRSHBERG

Dye installation at Tel Dor, Hellenistic period

walls and floors of their houses, possibly as insulation or simply as decorative elements.

The most important find relating to the dye industry at Dor was discovered on the southwestern corner overlooking the southern harbor in an area labeled "D" by archaeologists. Adjacent to the Persian Palace – actually a mansion from the Hellenistic period – they unearthed two large pits connected by a man-made channel dating back to that period. When first exposed, the pits, measuring about a meter deep and about the same in diameter, contained the residue of ancient purple dye. One of the pits was filled completely with broken murex shells, as Yisrael Hirshberg, the official photographer for the dig, recalls. "The pit was large enough to climb into, which of course I did. Even after we emptied it out, the sides of the walls remained embedded with broken shells, which I could easily identify as murex."[48] The other pit contained a large block of coagulated dye, dried and hardened over the thousands of years since it had been prepared. The remains at Dor are among the best preserved dye installations to have been found. There is no doubt that this was a dye factory of some sort, but scholars still debate the exact nature and specific function of the pits.

Stones lined the pit with the crushed murex. The second pit, however, had no waterproofing at all. When I first inspected the site, I formulated a theory that dyers collected the murex, broke them, and threw them into the first watertight pit. There the snail innards fermented over many days, perhaps heated by pipes carrying steam or hot water, as recorded in the recipes of the great naturalist Pliny the Elder and other ancient writers. When ready, the liquid would have been channeled into the second pit, where the dyers actually dyed the wool.

Ilan Sharon of the Hebrew University, the archaeologist leading the dig at Dor, however, invalidated the theory. The second pit, he said, since it was not watertight, could not have held dye long enough to color any wool. The current view among archaeologists is that the first pit held discarded shells, and the second, a sump, collected overflowing liquid. According to this reconstruction, the dyeing itself took place somewhere else, nearby. Possibly the pit acted as a receptacle in which a large jug was placed, or perhaps wool was dyed here by a method different from the one that Pliny describes. Instead of dipping the wool in a large vat of liquid dye solution, the dyers may have poured dye over the wool. The second pit would be filled with wool, over which fermented dye and then water were poured in repeating cycles until eventually the proper color appeared.[49]

In any event, the exact function of the pits remains open to speculation, and many questions still abide. These two pits alone couldn't have handled much traffic, nor would they have yielded industrial quantities of dyed wool, but the presence and size of the rectangular murex holding pens below the waterline along the coast indicate that the area saw the processing of huge amounts of murex. If so, though, where are the massive amounts of broken shells from all the years of dyeing? Where did the large-scale dyeing take place? To get a clearer picture of the dye industry at Dor, I turned to one of the key people involved in the excavation and exploration of the site.

With clear blue eyes, cropped silver beard, and a healthy tan, Kurt Raveh looks every bit the picture of a man who has spent most of his life at sea. But it's not sailing on the seas that interests him. He prefers searching for ruins and bounty lost underwater, exploring the secrets of wrecks buried beneath the waves. Born in Den Helder (Hell's Door), at the tip of the Dutch province of North Holland, Raveh grew up a *jutter,* the local term for beachcomber. After graduating from the naval academy there and serving in the Dutch navy in the Amazon and Caribbean, among other locations, he found himself in Israel. There Raveh continued to do what he loved best: scuba diving.

The Mediterranean Sea, it turned out, possesses archaeological treasures as rich as those of the Holy Land. Raveh became not only a maritime archaeologist, but also a leading expert in the field, writing books and articles and producing documentary films with National Geographic. He has worked on over forty major wreck sites around the world, and he helped establish the Center for Maritime and Regional Archaeology of Israel. But he focuses his research on the place he calls home: the beach at Dor.

Raveh agrees that the two ancient pits at Dor couldn't have served as the sole installation for the very large-scale dyeing that must have taken place given the size of the murex holding pens. He believes that industrial-scale dyeing took place in another location a bit north of the tell, past a beach that the locals call Love Bay, where archaeologists found the remains of shallow basins cut into the rock along with an aqueduct to bring in fresh water. Though all traces of dye have washed away over the intervening centuries, these interconnected pools could have produced huge quantities of dyed wool.

Raveh can also account for the missing shells. Up until the 1940s huge mounds of broken shells still littered Love Bay, the remains of all those years of dyeing. But in 1948 came Nahsholim, a kibbutz that still exists today. The farmers of Nahsholim harvested the shells, ground them up, and used them as an additive in their chicken feed, which improved the quality of the eggshells – at the expense of an important archaeological resource. Another possibility is that, even in ancient times, the shell middens were burned for quicklime. Mollusk shells consist almost entirely of pure calcium carbonate, and the kurkar around Dor provides a poor source for lime, used to make plaster and cement. It's possible that the lime found with the dye residue in the area D installation reveals a frugal recycling of the murex themselves.

According to Raveh, the two pits on the tell must have served as a boutique dye workshop. Indeed, the whole area facing the southern harbor seems to have been used for light industry. "That area looks like prime real estate with a gorgeous view of the ocean and a cool sea breeze," Raveh said. "Why would they have set up the industrial zone there and the residences inland instead of the other way round?" When he toured the site with a friend, a baker, Raveh realized that another consideration may have come into play. The baker explained that most ancient industries required furnaces or ovens, which in turn needed strong, stiff winds, which the sea would have provided, to keep fires burning.

Whatever the reason for the workshop's position, it wasn't the only time that folk wisdom augmented the book learning of academics and professionals. One of Raveh's most exciting finds was a ship in the Sea of Galilee dating to the first century, which became famous as "the Jesus Boat." Researchers encountered numerous problems and challenges in raising, transporting, and preserving the vessel. At one point, they found that worms had infested the water in the museum pool in which the boat was being kept, eating the wood of the ancient ship. Specialists around the world frantically tried to determine which chemicals might kill the worms without harming the fragile wood, but no one could provide a definitive solution. An old Arab fisherman came to the rescue, though, telling Raveh that he could take care of the problem. The fisherman arrived at the museum with two buckets of goldfish. Sure enough, the goldfish ate the worms and left the ancient wood unscathed.

On the winter morning that my wife and I sat in Kurt Raveh's beachfront house at Dor, sipping coffee, winds from the gray sky whipped the sand while the waves of the gray sea crashed upon the shore. Winter offers an important opportunity for underwater archaeology. After a storm, when seafloor sands have shifted, divers go out to see if anything has been uncovered. They mark sites that look promising and return for preliminary exploration in more favorable springtime weather. They choose one or two spots for detailed excavation, which then takes place in the late summer when the waters are at their calmest.

Marine archaeology presents all kinds of logistic challenges, regardless of weather or season, and requires creative problem solving. Frosted Mylar slates and graphite pencils allow for underwater writing. A tethered balloon can easily flag an important find; I have used balloons like that to mark the location of the baited traps we set to attract murex snails. A tube containing an air bubble can act as a makeshift spirit level. Dredgers – large underwater vacuum cleaners – suck up sand, which archaeologists then hand sift in the hope of finding ancient coins, jewelry, trinkets, or other artifacts. Raveh showed us one such item found in the hull of an ancient ship that sank off the coast of Dor: a two-thousand-year-old hazelnut that looked as fresh as if it had been picked from the tree yesterday. The fine sand along the southeast Mediterranean forms an excellent seal around objects that sink into it, making for exceptional preservation conditions.

Divers and snorkelers exploring the waters around Dor often come across barnacle-covered anchors strewn about the seabed. In the southern harbor alone, over 450 anchors have come to light, though that doesn't mean that all those ships met with ruin. A shrewd captain might opt to cut loose a heavy anchor and make up the weight difference with ballast of valuable merchandise. An anchor could easily be replaced at the port of destination, and some captains would risk sailing without one for the chance of making a greater fortune.

But of course many ships didn't reach safe haven. For thousands of years vessels have sailed along the eastern coast of the Mediterranean, the primary lane for transporting precious cargo among African ports, the Levant, and Anatolia. Just one ship a year sunk near shore by the strong westerly winds or storms would add up to thousands of wrecks waiting to be excavated.

Archaeologists have found only a very few so far, which means there are a lot more still out there waiting to be discovered.

At one point during our conversation, Raveh picked up a small package. Unwrapping a flannel cloth, he showed us a magnificently carved wooden statue of a woman, about a foot high. This, he said, was the goddess Tanit. "I found literally hundreds of these on a sunken boat off the coast a few miles north of here. Most were only a couple of inches tall, but some were nearly my height." The figurines, carved in Carthage 2,700 years ago, were making their way to Tyre for sale, but they never arrived at their destination.

Raveh has excavated dozens of dive wrecks all over the world, trained countless students, and done much to advance the discipline of marine archaeology, but he still dreams of unlocking from history a secret off the shores of Dor.

"Since my childhood days in Den Helder, whose beaches were actually visited once by Napoleon, I have been fascinated by the man. I am, in fact, a member of the secret Bonaparte Society," he furtively explained, showing us a medallion to prove it. "I have carefully studied the French general's memoirs, especially the parts relating to the time he spent at Dor."

Bonaparte had two very special cannons with him that fateful night in 1799. Those siege guns – the largest in the world at that time – were forged from the bronze of captured enemy weapons and decorated with scenes of victorious French battles. The pride of France, they signified the myth of her invincibility, and, wrote Napoleon's aide Louis de Bourrienne, "they made Europe tremble."[50] When Bonaparte jettisoned the artillery at Dor, leaving those two cannons behind dealt a devastating blow not only to the general himself, but also to the entire army. It drove home the fact that the war in Palestine was over and that France had met with defeat.

According to Raveh, men loaded the cannons onto makeshift rafts that horses pulled out to sea, but the rafts cracked under the weight of the heavy guns, and the horses, mired in the shifting sands, drowned.[51] As waves rolled over the mighty cannons, they sank into the sand and finally disappeared from sight. The pain and shame of losing those magnificent weapons was devastating to the men. "The soldiers seemed to forget their own sufferings, plunged in grief at the loss of their bronze guns," recounts Bourrienne. [52]

"I know those cannons are here somewhere, buried under the sand of the beach that I walk on every day." Raveh paused. "Perhaps one day I will find them."

V
TRUE BLUE

$\mathscr{T}$he Old City of Jerusalem, holy to people around the world, consists of four quarters: Christian, Muslim, Armenian, and Jewish. From the stone steps of the Jewish Quarter, you can see the Temple Mount, the site where King Solomon built the first Temple in the tenth century BCE. This sacred place stood for nearly four centuries until the Babylonians destroyed it and dispersed the Israelites into exile.

One hundred years later, the Persian kings Cyrus and Darius permitted the rebuilding of the Temple, but the new structure, built in the days of Ezra and Nehemia, proved rather small and unimpressive. In the last decades of the first century BCE, King Herod the Great undertook a comprehensive renovation and expansion campaign. His decision to construct a spectacular center of worship came partly as a matter of ego and partly as an attempt to ingratiate himself with the religious leaders of the time, who were not always enamored of him. Foreign architects and specialists designed the elaborate plans, but Herod involved the Temple priests themselves in the actual construction.

The completed edifice, the magnificent Second Temple, formed the centerpiece of a massive religious compound, a wonder of the ancient world. For all its splendor, though, it lasted only until the year 70 CE, when the Roman army under the command of Emperor Vespasian's son Titus destroyed it in the Siege of Jerusalem during the Great Jewish Revolt.

A truly majestic edifice, Herod's Temple consisted of the finest and most costly materials, including imported white marble that gleamed brilliantly in the sun. The Temple complex also comprised areas of increasing degrees of holiness and consequently increasing degrees of stringency regarding ritual purity. On the spot where the golden Dome of the Rock stands today, the innermost sanctum of Herod's Temple once stood. The Holy of Holies, as it was called, replicated the inner sanctum of Solomon's Temple, which had contained only one item, the famous Ark of the Covenant, containing the two tablets of the Ten Commandments. That ark disappeared when the Babylonians destroyed Solomon's Temple, so the new sacred space did not contain anything, just a raised platform where the ark had been. Only once a year did someone enter the Holy of Holies – the high priest, or *kohen gadol,* alone, on Yom Kippur – and only after he had purified himself fully and donned the required priestly garb. On that high holy day, he wore simple garments of white linen, but all year long the high priest's attire consisted primarily of *tekhelet.*

Leading to that sacred space was the Court of the Priests, to which all ritually pure priests had access. Here they walked up a ramp to the huge stone altar and offered the daily sacrifices of bullocks, sheep, doves, or simple grains on behalf of the kingdom. Laymen and laywomen could enter only the outermost section of the Temple, provided that they were not unclean – meaning that they had not recently come into contact with a corpse and had first immersed themselves in a ritual bath.

A supporting wall, its foundation buried seven meters into the ground and its height reaching twenty meters above ground level, buttressed the entire Temple compound – close to 150,000 square meters. In a remarkable feat of engineering, the wall consisted of huge stone blocks quarried locally that weighed tens and even hundreds of tons. Parts of this retaining wall still remain, particularly to the south and west; the western wall in particular became famous as the Wailing Wall, where people from all over the world still pour out their hearts in prayer.

The southern steps, excavated in the 1960s and 1970s, reveal an asymmetrical and staggered flight, designed deliberately to ensure a respectful, solemn gait. On the western wall adjacent to these southern excavations stands an outcropping known as Robinson's Arch, named for the great American biblical scholar Edward Robinson, who mounted an expedition to the Holy Land in 1838. During that expedition he also rediscovered for the Western world the famous fortress of Masada, another grand building project of King Herod. Robinson's arch, at the time one of the highest in the world, supported the upper part of a stairway leading from street level to the Temple above. The streets outside the Temple, noisy and crowded, bustled with people on their way to offer sacrifices and utter prayers, while eager merchants hawked their wares to passersby. It was atop these steps that Jesus confronted the money changers, overturning their tables that he felt affronted the sanctity of the holy Temple.

A bit north of Robinson's Arch stands Wilson's Arch, named for British explorer Charles Wilson, who rediscovered it for the West in 1864. One of a number of arches that supported a long overpass from the upper city, in what is now the Jewish Quarter, to the Temple, Wilson's Arch was used exclusively by the priests as they walked from their homes in the upper city and on Mount Zion to perform their tasks in the Temple.

Just a few steps from where this ancient road started, a door stands today with a simple sign: TEMPLE INSTITUTE. Jewish prayers since the destruction of the Second Temple have included a longing for the coming of the Messiah and the related hope that the Temple may be rebuilt "speedily in our days." The staff of the Temple Institute don't want to be caught unprepared. They have researched the historical, archaeological, and sacramental aspects of the fulfillment of this hope and are producing the necessary equipment for Temple services.

Those various day-to-day services and rituals required a plethora of tools, vessels, and other accessories. Expert craftsmen have designed and skillfully fashioned the giant copper laver for washing hands, the silver decanter for wine libations, the incense shovels to burn spices, and harps, lyres, and trumpets to accompany the singing of daily hymns. On display at the institute is the enormous seven-branched candelabrum, the Menorah of the Temple, standing taller than a person, created from one talent (around 130 pounds) of solid gold, as well as the golden Table of the Showbread.

In addition to creating the many metal vessels and tools, artisans at the Temple Institute have re-created the sacred vestments of the high priest, as well as the garments of his fellow priests. They fashion these garments from the most precious materials, including gold, fine silk, linen, and red wool colored with dye produced today from kermes insects found in Turkey. Besides red wool, the priestly clothing also consists of purple and sky blue wool – *argamman* and *tekhelet* – made of the dyes obtained from murex snails. To the high priest's splendid robe, made completely from *tekhelet,* and worn over crisp white linen pants and top, they have added a fringe hemmed with seventy-two pomegranate-shaped decorations alternating with seventy-two golden bells. In ancient times the chimes of those bells accompanied and announced the priest's every stride.

The much less ornate uniform of the lower priests, made primarily from white woven flax, also included an article made of pure *tekhelet* wool – a sixteen-meter sash wrapped around the waist. The Temple Institute has made 120 full sets of these garments that hang today in the closets of Jews of priestly lineage around the world. Those priests dream of the day when they will don the uniforms to perform their service in the Temple.

While some people dream of the future, others are still trying to understand the past. When archaeologist Shimon Gibson invited me to discuss his most recent discovery, he certainly piqued my curiosity. He was leading an important excavation on Mount Zion, the westernmost section of Jerusalem in the Second Temple period, and wanted my opinion regarding something he had found. The address he gave me was a modest building in a quiet Jerusalem neighborhood. As soon as I opened the front door, however, I realized that this was no ordinary apartment.

On every wall, from floor to ceiling, shelves bulged under the weight of countless bags of potsherds and other fragments.

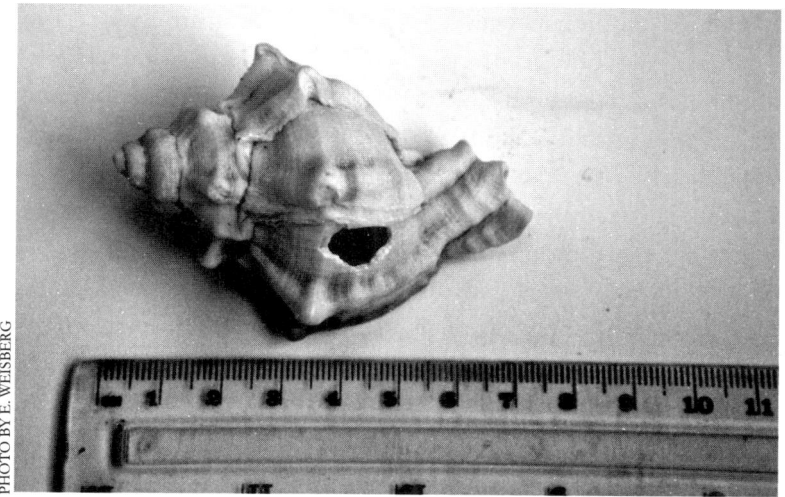

Murex shell (first century) found on Mount Zion

This storeroom held the finds from the Mount Zion dig, where researchers meticulously catalogued and stored the collected material, no matter how big or small, with the expectation that scholars one day will study the items and glean important historical lessons from some broken piece or weathered artifact.

Gibson took down one bag from the unending array and emptied its contents onto the kitchen table. What spilled out were not broken bits of pottery but rather shells – which I immediately recognized as *Murex trunculus*. Gibson explained that the dig had unearthed a few dozen of them, scattered around the Mount Zion site, and they dated to the first century, when the Temple of Herod still stood. The area, Gibson told me, situated close to the Temple, had served as a housing complex for priests and their families. In support of his assertion that this site once housed priests, Gibson had uncovered a ritual bath in the area as well as a beautiful stone cup with a cryptic ten-line inscription carved into it, presumably part of a purification ceremony. But what possible purpose, the archaeologist asked, could these shells have played here, and how did they make their way up the mountains to Jerusalem, a hundred miles from their habitat in the Mediterranean? The Temple enterprise required enormous amounts of murex dye to produce *tekhelet* for the clothes of all the priests and for the wall coverings and other decorative fabrics used in the building. Could the priests have brought the snails up from the coast to carry out the dyeing in Jerusalem? Gibson asked. It seemed unlikely.

Such an undertaking would have required transporting vast quantities of snails and keeping them alive for the duration of the journey since the dye can be extracted only from live animals. Further, the shell specimens he showed me from the dig remained whole. If they had been used for dyeing, they would have exhibited telltale breakages testifying to the extraction of the dye glands. Lastly, the smell that resulted from dyeing proved almost unbearable, so the process usually took place far from the center of a city. No, the dyeing would have taken place close to the sea, where the snails could be stored alive until needed; where necessary ingredients, such as salt, occurred in abundance; and where no one but dyers would have to abide the stench.

Gibson offered another thought. Perhaps the shells served a more mundane purpose: advertising. Since the Temple would have been a primary consumer of the blue dye, a priest or priests would have bought or arranged to buy the colored wool. They would travel to the coast and bring back the valuable material to be processed and spun in Jerusalem or nearby before being fashioned into the fabrics needed in the Temple. As merchants dealing in *tekhelet* for Temple purposes, they no doubt would have provided the blue strings of the tzitzit worn by the general public for religious purposes. Perhaps, as a striped barbershop pole advertises haircutting services, a snail shell outside a door declared to the public, "*tekhelet* sold here."

Priests provided *tekhelet* strings for the tzitzit of the population at large, but, I suggested, the shells' function would have been one not of marketing but rather of authorization. The correct analogy was not a sign announcing a service but rather a certificate of authenticity, conveying to the customer that he was getting the genuine article. Ancient buyers indeed had reason to beware, and they would have demanded proof that the *tekhelet* being sold came, as it properly should, from a murex snail. Then, as now, there was a trade in fraudulent religious objects – in this case counterfeit dye that unscrupulous merchants tried to pass off as authentic *tekhelet*. Cautious customers wanted to know that they weren't paying premium prices for fake blue strings known as *kala ilan*.

"The Holy One, blessed be He, will exact vengeance from him who attaches to his garment threads dyed with *kala ilan* and maintains that they are genuine *tekhelet*."[53] The Talmud uses this ominous threat to intimidate the unprincipled dealers who sold the counterfeit blue dye. Rabbis frowned upon dishonest behavior, of course, but this particular fraud could lead to overt violations of two biblical laws.

The first, from the book of Numbers, we know: that all the children of Israel affix the thread of blue to the corners of their garments; the second: "You shall not wear cloth combining wool and linen."[54] The one exception applies to the *tekhelet* strings of tzitzit. These blue woolen strings could be attached to a linen garment, and that was the common practice. The law is clear on both counts; only *tekhelet* obtained from the *hillazon* marine creature qualifies for ritual use in the tzitzit, and the only wool that can be worn with linen is the *tekhelet* strings of the tzitzit.

But another holy connection came into play as well. "Rabbi Meir said: Whoever observes the commandment of tzitzit, it is considered as if he greeted the Divine Presence, for *tekhelet* resembles the sea, and the sea resembles the sky, and the sky resembles God's holy throne."[55] The connection in the Midrash goes beyond mere color and may give us insight into the underlying reasoning for the rejection of *kala ilan,* even though it was identical in color to *tekhelet.* The statement insists that dye is meant to evoke the sea, and both its color as well as the creature that produced it, the *hillazon,* are of the sea.

Kala ilan was in fact plant-derived indigo, identical in almost every way to the blue dye obtained from the murex snails. Therefore substitution would go unnoticed. Since it was far cheaper to produce indigo blue than shellfish blue, the unscrupulous were inevitably tempted to use the fake. There was nothing wrong per se with selling indigo dye or indigo-dyed fabrics, of course, but the rabbis of the Talmud took seriously the religious implications of pretending that it was legitimate. "*Tekhelet* can only be purchased from an expert," they admonished, and they instituted precautionary laws requiring the transport of genuine strings under two seals.[56] In order to discourage further deception, they leveled divine curses against perpetrators.

Indigo made its way to the Mediterranean from China and India, and its name derives from the latter. It comes from a plant of the pea family, *Indigofera tinctoria.* Another form, woad, cultivated primarily in Europe from the flower *Isatis tinctoria,* yields far less dye, but the color obtained matches that of its Asian cousin – and *tekhelet.* Fashion and religion have different rules, though, and the blue from those plant sources, though identical to the eye, lacked sacred authenticity. The Roman elite had the same bias as the rabbis of the Talmud; the price of true snail-blue fabric far exceeded fabric dyed with the plant extractions. Further-

PHOTO BY ELWYN | 123RF.COM

Lan, *the Chinese character for indigo/blue*

more, when the Roman emperors regulated the manufacture and legal use of the shellfish dyes, they did not constrain the use of indigo in any way.

The origin of the phrase *kala ilan* is not completely under-stood, and several different explanations have been suggested. *Kala* in Sanskrit can mean blue, deep blue, or black. The second word, *ilan,* means plant or tree in Hebrew, and some believe that the phrase indicates the vegetable nature specifically of indigo dye. Another option connects the phrase with the Sanskrit *nilam,* sim-ilar to the Arabic *nil,* a term used for indigo. Rabbi Isaac Herzog, in his important work on the subject of *tekhelet,* suggests that per-haps the phrase comes from the Chinese word for the dye, *lan.* The Chinese character for *lan* consists of three elements – an eye, a person, and a vessel with water. Gösta Sandberg, an authority on indigo and author of books on the subject, notes that the an-cient Chinese term was actually *k'lan,* even more similar to *kala ilan,* but probably just a coincidence.

"*Tekhelet* cannot be tested," asserts the Talmud, meaning that, once applied, there was no way to distinguish the authentic dye from its counterfeit. [57] Some rabbis, however, maintain that chemical methods existed that could differentiate between the two. The Talmud records the formula of one rabbi who created a concoction of alum (aluminum potassium sulfate, used in ancient times as an astringent), sap of fenugreek (a white-flowered herb that gives curry its kick), and forty-day-old urine. He then soaked the blue strings in that mixture overnight to see if they faded. If they did, they were false. But that test didn't spell the end of the chemical adventure. Even genuine *tekhelet* will fade when placed in that caustic mixture. So a second test was devised to evaluate the faded string. It was baked in dough of barley bread to see if it regained its color – in which case it was deemed to be genuine.[58]

The pernicious use of indigo as fake *tekhelet*, though reprehensible morally, plays an important role from a scholarly perspective. Since archaeologists have discovered no ancient *tekhelet* fringes, and no ancient priestly costumes survive, we have no direct, firsthand evidence of the dye's actual hue.[59] Traditional sources fill in the gap, offering countersuggestions ranging from leek green to purple.[60] However, knowing that indigo could substitute, completely indistinguishably, for *tekhelet* corroborates the accepted traditional view that *tekhelet* was sky blue.[61]

Given that wools colored with authentic and fake dyes looked exactly the same, a savvy buyer naturally wanted proof of authenticity. The enterprising merchant-priest, looking to sell his strings, showed his customers that the *tekhelet* in his store was authentic by means of displaying *Murex trunculus* shells. Shimon Gibson found some of those shells – scattered in the aftermath of the Great Jewish Revolt that led to the razing of the Temple and the destruction of Jerusalem – buried there for two millennia.

VI
THE MISSING SHADE OF BLUE

$\mathcal{I}$n the 1960s, Israel's foremost archaeologist, Yigael Yadin, carried out detailed excavations at the desert fortress of Masada. A war hero, he began his military career at the age of fifteen and finished it twenty years later as chief of staff, the army's highest rank, in 1952. Then he traded his sword for a plowshare – or at least his gun for a shovel – and began a new life as an archaeologist.

The artifacts, bones, and structures uncovered at Masada tell the story of its history as a place of refuge and sanctuary for the persecuted, the pursued, and the paranoid. Prominent among the last was Masada's builder, King Herod, who made a habit of finding the most secluded spots in Israel to build his palaces. Masada was one of Herod's outstanding architectural achievements, a luxurious, Roman-style palace complex in the middle of nowhere, situated atop an almost inaccessible plateau on a rugged, barren mountain, overlooking the Dead Sea in the Judean desert. One narrow path snakes its way up a sheer cliff to reach it, and it may have been the only place where the distrustful king felt secure.

In later years, Masada served as the last stand of Jewish rebels who fled there during the Great Revolt against Rome. Three

The palace-fortress of Masada

years after Titus destroyed the Temple, a remnant of insurgents managed to hold out against the Roman siege at Masada, until Lucius Flavius Silva and Rome's Tenth Legion constructed a giant 375-foot ramp up the side of the mountain to breach the walls. According to the historian Josephus, the 960 defenders of Masada chose suicide over capture, and that traumatic scene has been imprinted on the collective psyche of Jewish people everywhere ever since.

Half a century later during another – and, as it turned out, final – Jewish revolt, Simon bar Kochba and his fellow fighters made their way to the same desolate place. Rebelling against Emperor Hadrian's religious persecution, they took refuge in nearby regions of the Dead Sea, including two caves about ten miles north of Masada along the bed of the Hever River, located at the oasis of Ein Gedi. Yadin excavated those caves as well. One is known as the Cave of Letters for what Yadin found there, and the second is the Cave of Horrors, named for skeletons and skulls, remains of the rebels who died during the Roman siege.

THE MISSING SHADE OF BLUE

As he dug through sand and stone, Yadin found many important relics that chronicled the life and death of those fighters. The dry desert air and the seclusion of the site, which lay virtually undisturbed for millennia, yielded finds exceptionally well preserved. Among the many documents and artifacts discovered at Masada and in the caves was one curious package of purple dyed wool. "The bundle of wool was found in the Letters-skin... wrapped in a piece of woolen mantle with coloured bands.... This find is of the greatest importance," Yadin wrote.[62]

Assuming the wool to have been dyed Tyrian purple, he sent it to an American dye chemist, Sidney Edelstein, for expert analysis. Edelstein's results revealed a mixture of blue indigo and red kermes. Since none of the other wool fabrics found there showed signs of red kermes, and since the bundle of wool was tied and specially wrapped, Yadin concluded that this wool was intended to be woven into strings for the ritual tassels.

The archaeologist relied on two assumptions to arrive at this conclusion. First, as Edelstein had shown, this wool hadn't been dyed with genuine murex *tekhelet* but rather indigo-derived *kala ilan*. Yadin notes that "the use of imitations of Tyrian purple was most common, though guilt lay only with those who used imitations knowingly." The second implicit assumption, held by all secular scholars in the 1960s, maintained that the color of true *tekhelet* represented a shade of purple rather than blue.

Subsequent research into the wool found at the Hever cave led most archaeologists and historians to reject Yadin's assertion that these woolen strands were intended for the ritual tzitzit, and the general himself beat a hasty retreat on the matter, explaining that his theory had been merely a casual suggestion. Accidental discoveries, however, and even mistakes of this sort can have significant, unintended consequences. Yadin was wrong about the tzitzit, but, as sometimes happens, his mistake aroused further interest, in this case the interest of Sidney Edelstein, which eventually led to a crucial breakthrough in our understanding of shellfish dyeing.

Like the shards and fragments of the Mount Zion dig, archaeologists bagged, tagged, and stored the countless items found at Masada for future study. The rich trove disappeared into the drawers and shelves of the basement of the Rockefeller Museum in Jerusalem. Nearly half a century after Yadin's excavations, though, one tiny scrap of clothing from the ancient site finally found its way to a chemist's microscope. Professor Zvi Koren of the Shenkar College of Engineering and Design announced in 2011 that his chemical analysis showed conclusively that the dark blue embroidered stripe on a small bit of cloth from Masada had been colored with dye from a *Murex trunculus* snail. The cloth dated to Herodian times and may have once draped the shoulders of some noble lounging about the royal courtyard at Masada. Yigael Yadin, who died in 1984, never lived to find out that he had, in fact, discovered authentic *tekhelet* after all.

Naama Sukenik is a scion of the Israeli archeological aristocracy, which includes Yigael Yadin (who changed his name from Sukenik) and his father, Eleazar Sukenik, founder of the Hebrew University of Jerusalem's archeology department, who first proposed the widely accepted theory attributing the Dead Sea scrolls to the ascetic community of the Essenes. It did not come as a surprise that Naama decided to pursue a career in archeology, specifically in the chemical analysis of ancient artifacts. Her doctoral thesis, under Prof. Zohar Amar of the Department of the Land of Israel Studies at Bar Ilan University, consisted of analyzing 180 fragments of fabrics found in various caves in the Judean Desert. Among those was one small piece of fabric from a cave in the deep desert ravine known as Wadi Murabba'at.

Situated in the northern part of the Judean Desert, Nahal Dragot (Hebrew for steps or levels) is colloquially referred to as The Darajeh and affords some of the most challenging canyoning in Israel. The local Bedouins divide The Darajeh into three parts based on geological features, with the central portion, Wadi Murabba'at (literally "the valley of the squares"), named for the square shaped caves that line the canyon walls. It was here that, in 1952, the French Dominican priest turned explorer, Roland de Vaux, discovered that the caves had been inhabited in the Roman era. He uncovered 170 texts including letters from the leader of the Jewish Revolt against Rome, Simon Bar Kochba, indicating that a band of refugees loyal to Bar Kochba had used those caves to seek shelter. Sometime around the year 135 CE, bits of cloth from those refugees must have dropped to the cave floor, lost and forgotten until de Vaux or subsequent archeologists picked them up. In 2013 they made their way to Naama Sukenik's lab for analysis.

Of all the fabrics analyzed, Sukenik found that three were colored with dye obtained from Murex snails. Two of those were a reddish purple hue, a mixture of purple from the Murex and red from cochineal. The third, however, Textile No. 22, had a blueish stripe – a dye obtained solely from *Murex trunculus* snails. Furthermore, while the two purple fabrics were woven with a z-

PHOTOGRAPH BY YEDIDYA STERMAN

The caves of Wadi Murabba'at

twisted yarn typical of imported fabric, Textile 22 with the blue dye was woven with an s-twisted yarn, which may suggest that it had been manufactured locally in Israel.[63]

The Masada fabric that Yigael Yadin had found offered no context or provenance; one might argue that it had belonged to some Roman noble visiting the Herodian palace. Naama Sukenik's analysis of the Wadi Murabba'at textile, however, leaves absolutely no room for doubt that the ancient Israelites encountered the technology that transformed the purple Murex dye to blue, and may provide the first physical evidence of the manufacture of *tekhelet* locally in first/second century Israel.

The stripe of *tekhelet* found on the Herodian cloth and the Wadi Murabba'at fabric reflect a prominent cultural reality. In

Greek and Roman society, purple-dyed wool symbolized status, and status signified power. After Alexander the Great defeated Darius III and victoriously entered the Persian capital, Susa, the Macedonian conqueror found there about fifty-five pounds of purple wool dyed 180 years earlier that still retained its beautiful luster. When Alexander's men reached the tomb of Cyrus, they found next to the Persian king's coffin a couch standing on golden feet, on which "a carpet of Babylonian tapestry with purple rugs formed the bedding... and robes dyed the color of hyacinth* were also lying upon it, as well as others of purple and various other colors."[64]

To Alexander, purple and blue came to represent the glamorous, exotic world he had conquered; to the dismay of many of his soldiers, he began to adopt and institute the customs of the empire he had conquered, including the Asian practice of falling prostrate before a ruler and the wearing of luxurious royal-colored garments.

Clothes may or may not make the man, but in ancient Rome the toga came in a variety of styles for a variety of men, each with its own specific significance or indication of social status. The *toga virilis*, worn by the ordinary freeborn male citizen who had come of age, consisted of plain off-white fabric. The black *toga pulla* signified mourning. Seekers of political office wore a toga bleached white with chalk, the *toga candida* (whence the word *candidate*), to show the supposed purity of their motives. But the honor of the highest form of dress went to a triumphant general

---

* *Hyacinth* is the term the ancients used to describe what we call blue, and the ancient Greek translation of the Hebrew Bible, known as the Septuagint, translates the word *tekhelet* as *"iakintos"*— hyacinth.

returning from battle, the *toga picta,* completely purple and richly embroidered with gold. "It brightens every garment," writes Pliny the Elder, "and shares with gold the glory of the triumph. For these reasons, we must pardon the mad desire for purple."[65] Earlier kings had worn an all purple garment, the *toga traeba* or *toga purpura,* but Julius Caesar, emphasizing his military victories, began the practice of wearing the *toga picta* as his standard dress, and subsequent emperors followed suit.

Certain royal or notable figures associated with purple throughout history, however, would presumably have passed on that distinction. Suspicion has it that such famous people as Nebuchadnezzar, Mary Queen of Scots, Vincent Van Gogh, and England's King George III suffered from a genetic disease known as porphyria. The illness gets its name from the purple color of the patient's stool and urine, and is caused by a disruption in the production of heme, a component needed for healthy blood cells. In some cases the nervous system is affected, leading to seizures, hallucinations, and other mental disorders. Many suggest that the dementia suffered by the king of England and referred to as the 'madness of King George' was due to this disease rather than to grief over losing the American colonies.[66]

Exhibiting some of the autocratic behavior that later convinced Brutus and the other Liberators of the Roman Senate to assassinate him, Julius Caesar introduced restrictions regarding the use of purple clothing. Previously, throughout the ancient Near East shellfish dyes had proved fantastically expensive. But economic constraint rather than imperial fiat limited their use. "The use of the color purple was never, however, neither among the Persians, nor in fact in any other ancient society, interdicted to private persons. It was used widely as a sacerdotal and cultic color and by private individuals as a form of luxury display."[67]

In Greek and especially Roman times, however, this tolerant attitude began to change. According to Roman biographer Suetonius, Caesar allowed only "those of a designated position and age, and only on set days" to wear purple robes. Caesar's grandnephew and successor, the emperor Augustus, similarly objected and "gave orders that no one should wear the purple dress except senators acting as magistrates."[68] Last of the Julio-Claudian emperors, Nero even went so far as to set up a sting operation to catch violators of the laws. "Having forbidden the use of amethystine or Tyrian purple dyes, he secretly sent a man to sell a few ounces on a market day and then closed the shops of all the dealers" and confiscated their property.[69]

After the destruction of the Temple in Jerusalem, the priestly service came to an end and with it the manufacture of the blue cloth that the priests wore. Jews, however, both in Palestine and the rest of the fledgling Roman Empire, continued to follow the biblical commandment to wear a *tekhelet* string on the fringes of their garments. How edicts restricting the use of purple affected them remains unclear, particularly since "purple" represented such a range of colors in the ancient world, but *tekhelet,* mirroring indigo as it does, may have escaped the harsher scrutiny and regulation that befell what we know today as purple. Then again, maybe not. The judges who sat on the Sanhedrin, the Jewish high court, may have worn robes with a band of *tekhelet* similar to those of Roman senators, and perhaps their garb would have been subject to the laws against purple.[70]

Under Constantine – who in 313 CE issued the Edict of Milan, pronouncing religious tolerance throughout the empire – and his longest reigning son, Constantius II, officials slackly enforced the measures against wearing purple.[71] Yet the Talmud records a story of a pair of travelers from Tiberias, on the Sea of Galilee, who tried to smuggle *tekhelet* to the Jews of Babylonia, "and the eagle caught them."[72] The eagle traditionally represented Rome, but it could also stand for the Persian army, so it's not entirely clear whether the Romans were barring the export, or the Persians the import, of the dyed wool. Either way, dealing in *tekhelet* had become a dangerous business, and a beautiful Roman coin showing the imperial eagle with a murex between its talons evokes the important role that shellfish dye still played in the region in the late empire.

Third-century Tyrian coin
ARI GREENSPAN | TEKHELET.COM

Constantine, who founded Constantinople as the capital of the Eastern Roman Empire – eventually known as the Byzantine Empire – had his own factory in Tyre for the production of purple dye.[73] The enthusiastic admiration of Byzantine emperors for the color bordered on veneration. The association of purple with imperial status reached its greatest extreme with them, their magnificently decorated purple robes considered sacrosanct. Official state documents were written on purple parchment, and special purple ink was used for royal signatures. Rewarded for faithful service, officers of the state received the privilege of "adoring the sacred purple," in other words: appearing at court in order to pay homage to the emperor. An empress giving birth to a legal heir did so in a room specially decorated with purple marble, and the term *porphyrogennetos* (πορφυρογέννητος), literally "born to the purple," became an honorific title bestowed upon the child delivered there.

Emperor Theodosius II, in his fifth-century magnum opus, the *Codex Theodosianus*, compiled and streamlined all the laws and legal precepts issued in the previous century, when the empire had begun to Christianize in earnest. The project took twenty-two scholars over nine years to complete, and it was presented to the senates of Rome and Constantinople on December 25, 438. Theodosius outlawed not only the wearing of purple-dyed clothes; manufacturing and even owning the material also became a criminal act.

All persons of whatsoever sex, rank, skill, profession, or family shall abstain from the possession of that kind of material which is dedicated only to the Emperor and to His household... Men shall bring forth from their homes and deliver the tunics and cloaks that are dyed in all parts of their texture with the blood of the purple shellfish... Garments of all-purple must be surrendered to the treasury.... There is no reason why any man should complain of having been deprived of the price, because it shall suffice that he obtains impunity for the violation of the law that he trampled... nor is there any reason why he should have occasion to worry about profits, since his life does not have to be placed at stake. But let no man now by such a concealment incur the peril of the toils of the new constitution; otherwise he shall sustain the danger of involvement in a crime similar to high treason.[74]

Emperors spilled much legal ink – purple or otherwise – on Tyrian purple, but what of our blue-dyed wool? Titus had destroyed the Temple in 70 CE, just a few centuries earlier. Had the esteemed blue fabric already disappeared?

It had not.

Justinian I, one of the great emperors of the East, like Theodosius II collected imperial decrees before him. The *Codex Justinianus,* which appeared in 529, itself part of his *Corpus Juris Civilis* (Body of Civil Law), completely reworked and rewrote all of Roman law, shaping and influencing European law for many centuries. Book 4.40 of the *Codex* begins:

WHAT CANNOT BE SOLD AND PEOPLE WHO ARE FORBIDDEN TO SELL OR BUY.

No private citizen shall have the right to dye or sell purple cloth, either silk or wool, in the colors called blatta, oxyblatta, or hyacinthina. If anyone should sell the aforementioned dyed wool, let him know that he risks losing his property and his life.

We can take two very important pieces of information from this. First, purple cloth still possessed such importance that it headed the imperial agenda of commodities to regulate – a list that ran some sixty-six items long.[75] Second, and more importantly, is the presence of the word *hyacinthina,* which we've seen before. In the tomb of Cyrus, Alexander the Great had found "robes dyed the color of hyacinth." This law forbidding the sale of "purple" cloth tells us that for legal purposes *tekhelet* was considered purple, and therefore all the laws and interdictions about Tyrian purple – from Julius Caesar to Justinian, over more than five hundred years – also applied to *tekhelet.*

It further tells us that, for Jews throughout the empire, adhering to the commandment to wear *tekhelet* was becoming nearly impossible. Even if legally they could wear it, the impoverished community couldn't afford the precious strings. Imperial authorities strictly controlled the industry, making the material extremely difficult to obtain even for those who could pay. Add to that the life-threatening danger that wearing or even owning *tekhelet* could incur.

The rabbis of the Talmud understood the unbearable strain that the injunction in the book of Numbers entailed, so, while not renouncing the commandment, they offer words of comfort for those unable to fulfill it. "Greater is the punishment for [those who do not wear] white, than for [those who do not wear] *tekhelet*," since it's a simple matter to obtain white strings for the tzitzit, and therefore there is no excuse for failing to wear them. [76] *Tekhelet,* though, was another matter, and God would take into account the circumstances of someone who couldn't wear the cherished sky blue threads.

Hopeful that in the future the practice would flourish again, the sages carefully kept the tradition alive, studying and preserving the secrets of manufacturing the dye. One of the last recorded conversations of the great Babylonian scholar Abaye was with a recent traveler from the land of Israel, Shmuel son of Judah.

"This *tekhelet*," Abaye asks, "how do you dye it?"[77]

But efforts to preserve the practice and even to safeguard the details of the production process were proving essentially futile. The last mention of it in Talmudic times comes in a laconic sixth-century statement regarding a certain Mar of Mashkhi, who once obtained it. The earliest post-Talmudic works, dating to the eighth century, confirm that the practice has ended. "But now we have only white, for *tekhelet* has been hidden."[78]

It appears that the tradition of wearing the blue strings, as well as all knowledge of how they were produced and even the natural sources of the dye, disappeared for the Jews during the seventh century.[79] The most probable date for the end of the *tekhelet* industry in the land of Israel itself is the first half of the seventh century.

It was a disastrous period for the Jews in Palestine. In 614 the Persian Sassanid general Shaharbaraz conquered Jerusalem from the Byzantine Romans and triumphantly carried off the cross of the crucifixion. Emperor Heraclius regained control of the city for the Romans and restored the true cross in 629, only to lose the city conclusively to Caliph Umar in 638. The land of Israel and Jerusalem were conquered no fewer than six times in the space of just twenty years. When Arab armies took Palestine, they destroyed anything associated with Roman rule, including dye factories, which had been under imperial control. The destruction of the dye houses and the decimation of the Jewish community in the Holy Land put an end to the secret dyeing processes, handed down from father to son, and to the traditions regarding the biological identity of the *hillazon*. With the passage of time, the details grew more and more obscure.

To the rest of the world, the availability of the much cheaper plant-derived indigo made the loss of murex blue less significant. Even the creation of purple without shellfish didn't pose a problem since dyers could make it with indigo and perhaps kermes, mixed together, as Sidney Edelstein had seen with the piece of wool from the Bar Kochba revolt. Authentic murex purple still appeared sporadically as the Arabs chased Christendom around the Mediterranean. However, with the defeat of Emperor Constantine XI Palaeologus by Sultan Mehmed II and the fall of Constantinople in 1453, all traces of murex dyeing vanished. As dyers and merchants turned their attentions elsewhere, the secrets of the murex, how to find them and how to dye with them, slipped further into oblivion.

VII
MOOD INDIGO

$\mathcal{L}$ong ago, a young mother went down to the river to offer a sacrifice to the water spirits. On the mother's back, wrapped in a white cloth – for this was before man knew the secrets of dyeing, and all cloth was white – her infant daughter slept soundly. At the bank of the river the mother laid the child down on a pile of leaves and made a fire to cook a portion of rice for her offering.

In those days, the sky was still very close to the earth, especially near the rivers, where it was hard to tell where the earth ended and the sky began. When hungry, one could simply reach up and eat a piece of the sky. So enchanted was the mother by the beautiful blue color of the sky reflected in the water that she began to hunger for it. While the rice slowly cooked, she broke off a bit of the sky and ate. It was so tasty that she ate more and more until, feeling satisfied and drowsy, she fell asleep.

She awoke suddenly to the smell of burning rice and saw, to her great anguish, that while she had been sleeping her daughter had rolled off the leaves and was lying smothered by the surrounding tall grass. As she lifted the lifeless body, she noticed that the child had wet herself and that a patch of blue appeared on the white cloth where the baby had lain. The mother wept and wept until she fell into a dreamful slumber.

In her dream the water spirits explained that the leaves on which her baby had slept were indigo, and that for the blue of the sky to come down to earth and remain permanently, all that was needed was indigo leaves, urine, and ashes. The water spirits commanded her to teach the women of the village how to dye with indigo, and in return the women would forever bless the memory of the young mother who sacrificed her child and gave them the secret of bringing blue down to earth.[80]

This charming Liberian legend, like similar stories about the origin of shellfish dyeing, illustrates the recurring belief that the ability to produce something as marvelous as blue fabric can only have resulted from some kind of supernatural intervention.[81]

If indigo dyeing had been a simple matter of combining indigo leaves, urine, and ashes, or if murex dyeing had required just squeezing some liquid from a snail gland, it's doubtful that the resulting colors would have been considered so remarkable and desirable. It was the complicated, protracted processes, the seemingly magical transformations, the inexplicable changes of color from blue to yellow to green to the final blue that led to the high prices that artisans charged and that people were willing to pay. Creating blue from sticky bits of shellfish or hairy leaves and green shrubs seems an impossible undertaking. The odds are stacked against it, and the colors that rival the azure sky ought not to exist. But they do.

In the ancient Middle East, along the shores of the Mediterranean Sea where murex dyeing was born and where craftsmen over generations honed it into a fine art, it was specifically the shellfish dyes, pride of the region, that received adoration, deemed worthy of royalty and befitting the service of God. The Talmud clearly reflects this partiality when it rejects the use of *kala ilan* and prescribes tests, not always very effective, to distinguish between the imitation and the real thing. Also, as we saw, the strict edicts of the Roman emperors against appropriating blue- and purple-dyed fabrics for common use did not extend to indigo-derived dyes since fabrics colored in that manner conferred no superior status. No self-respecting emperor, empress, prince, or princess would ever deign to wear apparel made with such an inferior dye.

In the Far East and Africa, however, the situation was different; shellfish dyeing was not known there, and indigo was considered noblest of dyes, a gift of the gods, who in their generosity gave man the ability to touch a piece of the sky. In ancient India and China, some revered indigo for its aesthetic value and others for the profit it could bring to those who grew and traded it. For centuries indigo ranked in the West as the new prized blue, gaining the status that once had belonged exclusively to the shellfish dyes.

It should come as no surprise that legends such as the aforementioned Liberian one, portraying the discovery of dyeing with indigo as a sacred gift from the gods, span the globe and appear in various forms in many diverse cultures. Much about the origins of indigo dyeing – how, when, and where it all began – remains unknown; presumably the process arose by chance at different times in different places. What we also may never know is which preceded which, indigo or shellfish dyeing; by the second millennium BCE both methods were known. As its name indicates, indigo seems to have originated in India and spread from there across the globe, but we can't know for sure. In ancient Egypt, imported or not, indigo or woad was dyeing wool and linen. Among the magnificent finds from the tomb of Tutankhamun (around 1336 BCE) we have a small indigo-dyed kerchief that the young pharaoh might have worn as a child.

Julius Caesar famously wrote of the custom of ancient Briton warriors to paint themselves blue: "All Britons, in fact, stain themselves with woad, which produces a blue color, so that they are more terrifying to face in battle."[82] Caesar was talking about the Britons in southern England, though people often apply his observation to the Picts, a people living in what is now Scotland. Although we have little reliable evidence to corroborate the belief, the Picts – from the Latin *picti,* meaning "painted men," – daubed themselves with blue paint before battle. Anyone who has seen the movie *Braveheart,* set in Scotland long after the Picts disappeared, remembers the painted warriors from the battle scenes. If in fact this was a Pictish custom, they most likely used locally available woad to create the intimidating effect.

Better attested is the custom of elaborate decorative blue tattooing. Although some experts maintain that tattooing with woad is possible, others disagree. Pat Fish, a seasoned tattooist who claims descent from the Picts, described her experience at trying to practice her art with woad-based ink, which she called

an amazing astringent. The tattoo I did with it literally burned itself to the surface, causing me to drag the poor experimented-upon fellow to my doctor, who gave me a stern chastising for using inappropriate ink. It produced quite a bit of scar tissue, but healed very quickly, and no blue was left behind. This leads me to think it may have been used for closing battle wounds.[83]

Experiences similar to Pat Fish's may account for the belief in many cultures, both ancient and modern, in the healing powers of indigo and woad dyes. Thought to reduce fevers and function as a general antiseptic, indigo was rubbed directly onto the body as well as taken internally. A Hindu Siddha medicinal recipe mixes indigo with honey to help alleviate jaundice and diseases of the liver. In traditional Chinese medicine, indigo detoxifies the blood and reduces inflammation, while in South Africa it mollifies toothaches. Online you can buy hair oil from Kerala, India, made from coconut milk, gooseberries, and *Indigofera tinctoria,* which promises to reduce hair loss and aid in "growing long dark and lustrous hair." The disinfectant properties of indigo might explain why plant species produce the substance, which may serve as a natural insecticide or antibiotic protecting the plants against pests or blight.

The ancients didn't understand the underlying chemical processes, of course; all they knew were the appropriate quantities of the various ingredients and the sequence of steps that somehow resulted in the rare and valuable color. Given the lack of scientific understanding, all sorts of superstitions arose about the dyeing process and about the power of indigo. Some cultures, for example, believed that if a woman passed the dye vat at a certain stage in the dyeing process, she would spoil the dye. In other cultures it was specifically a menstruating woman who, it was thought, could ruin an entire crop of indigo plants if she walked past. On one Indonesian island, dyers hung chicken feathers above the dyeing site to ward off evil spirits, and on other islands chickens or piglets served as ritual sacrifices before any dyeing began.[84]

Many hundreds of plants produce indigo, the most important being *Indigofera tinctoria,* from the pea family, which grows best in tropical and subtropical climates, and *Isatis tinctoria,* a flowering plant related to cabbage and horseradish (more commonly

Indigofera tinctoria
© MORPHART CREATIONS INC. | SHUTTERSTOCK.COM

known as woad, or in French *pastel*) that grows in other regions, including Europe. Older plants are cultivated for their seeds to produce new plants since only the first-year crop of woad can provide blue for dyeing. Jenny Balfour-Paul describes the intricate and labor-intensive procedure for preparing indigo from its plant source, in her authoritative work on the dye:

> *The real challenge for farmers is extracting the dye from their leaves.... Processing indigo could be done in three ways: first the simple method, where fresh leaves were put directly in the dye pot; second, that where*

the leaf mass was processed and fermented but the dye pigment not extracted; and third, the most sophisticated method, where the indigo was fully extracted from the leaf mass.[85]

Dyers could compost and ferment woad leaves, forming them into balls for transport. But these leafy clumps were not fit for long-distance export; they were cumbersome and tended to mold. *Indigofera tinctoria,* on the other hand, could yield large volumes of transportable dye. In colonial India, indigo plantations supplying dye to Europe formed a pillar of the economy, and a complex traditional process that bordered on the ceremonial accomplished the extraction of the dyestuff. Covered with water, the branches fermented for a day and a half until a blue scum bubbled to the surface. The timing of the next stage, according to expert dyer Gösta Sandberg, was critical.

As soon as the liquid tasted sweet and was dark blue in colour, it was quickly drawn off from the vat into a tank on a lower level, where women and girls stood ready with long bamboo sticks in their hands. As soon as the tank was full they began stirring and beating the liquid. With rhythmic movements of the whole body they whipped the liquid until the entire surface was covered with a thick layer of scum which started by being blue, but became whiter and whiter towards the end and then disappeared altogether.[86]

This "whisking" went on for a few more hours until patches of dark blue formed. These blue clusters were then left to settle to the bottom and could be extracted and formed into cakes that slowly dried over two or three months. In this solid form, indigo was easily transportable and could travel, so that the dye became an important article of trade throughout the world.

In medieval Europe, imported indigo, a valued commodity, commanded expensive prices, whereas homegrown woad, generally considered inferior, was less expensive and used widely both as a blue dye and as a basis for many other colored dyes. Toward the end of the fifteenth century, when the ocean routes to India and China opened following Vasco da Gama's famous expedition, indigo became one of the chief commodities imported, along with such other exotic items as black pepper, nutmeg, and opium.

Over the next two centuries the oceans became crowded with ships from such newly formed corporations as the Dutch East India Company and its rival, the British East India Company, bringing vast amounts of indigo to Amsterdam and London, and creating an extremely lucrative trade. Protests erupted across Europe, however, as imported indigo posed a serious threat to the livelihoods of local woad dyers. Foreign indigo became known as "The Devil's Dye," and bans against it protected the local woad industry. In 1609 King Henry IV of France passed a law sentencing to death anyone who used "the false and pernicious Indian drug." But the superiority of *indigofera* over woad was undeniable, and even royal law couldn't suppress its use. Despite attempts to legislate against the trend, European woad growing and dyeing dwindled, replaced by imported indigo.

As the demand for indigo grew, the nations of Europe, now colonial powers, turned to their colonies to grow the valuable plant. South Carolina, for instance, became Great Britain's biggest producer of indigo once cultivators discovered that indigo could successfully grow there, and it rapidly became the region's most important cash crop after rice.

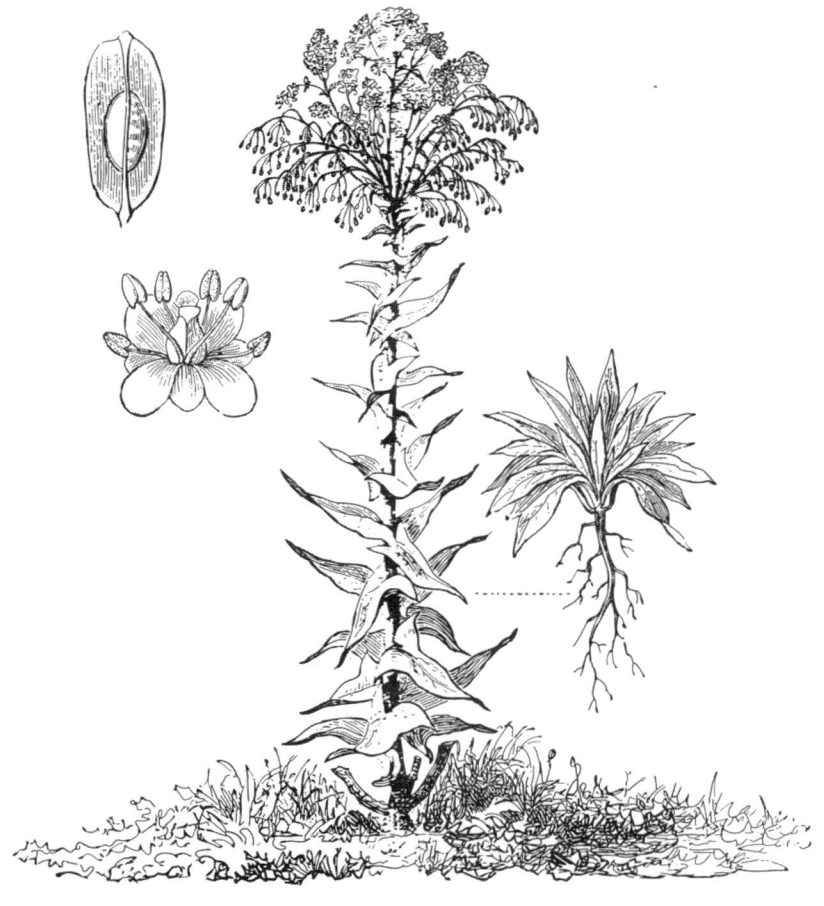

Isatis tinctoria
© HEIN NOUWENS. | SHUTTERSTOCK.COM

Credit for introducing indigo to South Carolina goes to a remarkable young woman by the name of Eliza Lucas Pinckney. Born in Antigua, British West Indies, and educated in London,

Eliza moved to South Carolina with her family in 1738. Soon afterward her father, a lieutenant colonel in the British Army, returned to Antigua, leaving sixteen-year-old Eliza in charge of the family, their three plantations, and the plantations' many slaves. Eliza had received the traditional female education of the time, focused on respectable, indoor subjects such as French and music, but she had developed a special love for botany.

When her father sent indigo seeds from the Caribbean, she eagerly experimented with them until, after several failed attempts, she finally grew them in the new climate. This intelligent, independent young woman shared her knowledge with other farmers, and within two years the volume of indigo exported from the Charleston area increased over twenty-five-fold from 5,000 pounds to an astonishing 130,000, bringing great wealth to many plantation owners. She experimented with other crops as well, including figs, hemp, and silk, and met with so much success that she became the first woman inducted into South Carolina's Business Hall of Fame. When she died in 1793 at the age of seventy-one, President George Washington served as one of her pallbearers.

The enormous popularity of indigo came from its rich and enduring color, a result of its unusual chemistry. Plant indigo and murex-based dyes, both of them variations of the indigotin molecule, have distinct advantages over other natural dyes. In order for other dyes to adhere to a fabric, they require the aid of a mordant, an intermediary chemical, usually iron or alum, which helps fix the dye to the fibers to form a colorfast fabric. The process for coloring cloth with murex and indigo is known as "vat dyeing," and the nature of their indigotin chemistry eliminates the need for a mordant. In vat dyes, the chemicals in their natural state don't dissolve in water and cannot adhere to fabric. They must undergo a chemical transition called reduction, which allows them to dissolve in water and to be introduced into the fiber matrix. In earlier times, urine facilitated the reduction. Nowadays, other reducing agents such as sodium dithionite do the job. During this stage the dye solution appears a green-yellow color. Removed from the vat and oxidized upon exposure to air, the dye on the cloth returns to its original color, and the fabric transforms from green-yellow to a perfect blue or purple.

Because of this uncommon chemical process, indigo, unlike most other natural dyes, can color a broad array of different fabrics, including cotton, flax, and silk, though it binds particularly well to wool. As we know from archaeological finds, the color, like that obtained from murex, can remain vibrant for centuries.

In small-scale dyeing, which many enthusiasts undertake today, it's possible to gather the ingredients in a pot and allow the chemical reactions to take place. For larger-scale production, however, each stage in the process requires meticulous monitoring. If the leaves aren't properly handled, if fermentation goes just a little too long, if the temperature of the liquid isn't right, or if the dyeing begins too soon – any one of these slips can affect the color and stability of the dye, more often than not negatively. Experts either had to carry out, or at least supervise, the entire process for satisfactory results. The alkalinity of the dye vat played a vital role, and without the pH scale – not introduced until 1909 – intuition more than precise measurement kept that parameter within its proper bounds.

In his 1806 work, *The Dyer's Companion,* eminent American dye master Elijah Bemiss describes three methods for testing the alkalinity. First, you can rub the liquid between your fingers, a slippery feel indicating more alkalinity, a rough feel greater acidity. Second, an expert dyer can distinguish subtle differences in smell of a more or less alkaline vat. Third, after a small sample of cloth has been dipped in the dye, the resulting color will be dull or bright if the solution is more or less alkaline.[87] John Edmonds, an expert in medieval dye techniques, quotes a further method mentioned in the literature: "by taste on the tongue giving a pungent taste."[88]

As indigo became more popular and trade in the dye became more lucrative, a race began to produce it artificially. In 1878 German chemist Adolf von Baeyer first synthesized indigo and received the 1905 Nobel Prize in chemistry for this important discovery and his groundbreaking research into other organic dyestuffs. But Baeyer's methods proved impractical and cost-inefficient. In 1880, before BASF, the Baden Aniline and Soda Factory, became the industrial giant it is today, the company mounted a massive seventeen-year project at a cost of eighteen million gold marks to mass-produce indigo. It finally brought synthetic indigo to market in 1897, due largely to an accident.

The breakthrough had to do with one particularly expensive stage of the process in which naphthalene, a substance found in tar, boils to create phthallic anhydride (PA). One day, while measuring the temperature of the vat, a chemist named Eugene Sapper dropped his thermometer, which broke and spilled mercury into the mixture. This unexpected addition increased the efficiency of PA production immensely and in turn greatly increased operational effectiveness.

That broken thermometer had far-reaching political ramifications as well. BASF quickly cornered the indigo market with its cheaper, pure synthetic product. With the plentiful availability of a cheaper dye, European industries no longer needed natural indigo, and the nations gradually stopped importing it from their colonies.

The indigo of the British Empire grew on colonial plantations in India, founded in the eighteenth century. The ruthless exploitation of farmers by zamindars – wealthy hereditary landowners who forced farmers to plant indigo rather than food crops – led to a revolt in Bengal in 1858. More than half a century later, in the Champaran district of Bihar in northeast India, the aristocratic zamindar class passed losses from the plummeting natural indigo market to the farmers, driving them into debt and poverty, causing another uprising.

Riots broke out in 1914 and 1916. Mahatma Gandhi visited Bihar a few years later and saw firsthand the plight of the indigo farmers. Demands that the colonial government offer relief from the heavy tax burden met with inaction, so it was in Champaran that Gandhi initiated his first satyagraha, a nonviolent protest of mass civil disobedience. The British promptly arrested him, but his method proved a formidable form of dissent, eventually overpowering English colonialism and helping to bring independence to India.

Meanwhile, synthetic indigo turned out to be less profitable for BASF than they had hoped. Another BASF product, indanthrene blue, also known as carbon paper blue, was replacing indigo as the blue dye of choice for fabrics and clothes. But just as in the old Westerns that had begun flickering on silver screens, cowboys came riding in to save the day. Levi Strauss – a dry goods seller who had emigrated from Bavaria to New York and then, following the discovery of gold in Northern California, to San Francisco – and Jacob Davis, his business partner from Reno, Nevada, invented a particularly sturdy version of workman's pants that we know now as denim jeans. American teenagers adopted the rugged trousers in the 1950s – in everybody's favorite color, blue – and adults soon followed suit. Because indigo doesn't bond well to the cotton in denim, it fades over time, giving jeans the well-worn look that continues to swing in and out of popularity in modern fashion. More than one billion pairs of jeans are manufactured worldwide each year, firmly establishing indigo as the king of dyes.

The developments over the centuries that led to more efficient indigo production and later to the whole range of synthetic blue dyes had no bearing on the ritual *tekhelet*. Only the authentic blue dye obtained from the *hillazon* could legitimately be used for the tzitzit. Rather than use what they considered a counterfeit dye – albeit indistinguishable from the genuine article – Jews continued for many generations to wear only plain white tzitzit.

But the status quo was about to change.

VIII
THE QUEST FOR THE HOLY SNAIL

The young scholar had been studying in Lublin for only a few months, but already he was growing restive. His teacher, a Hasidic master known as the Seer of Lublin, was a great mystic who it was said could see from one end of the world to the other and even probe into the very essence of a man's soul. The old rabbi had sent emissaries to find the young prodigy and bring him to his court.

At their very first meeting, the clairvoyant sage had accurately identified the contents of the boy's pockets.

"I came here to learn from your wisdom, not to see miracles," said the brash youngster, unfazed and unimpressed.

The relationship between the two did not improve, and the student soon decided that Lublin was not the place to reach his spiritual potential and grow in knowledge and piety. His mind made up, he left for the town of Pshischa to learn from a different Hasidic master, called the Holy Jew. With no inclination for social graces such as polite good-byes, the boy didn't even bother taking his leave of the Seer.

But walking down a path that led away from the study hall, he heard a voice. Was it just another student calling to him? Was it perhaps a heavenly whisper? Or was it his own conscience echoing in his head? "As you, Menachem Mendel, left your teacher abruptly and broke his heart, so, too, your greatest student one day will leave you."

The 1700s saw a new movement arise among the Jews of eastern Europe. It came to be known as Hasidism, and it reacted, among other things, against the prevailing excessive legalism that regarded the academic mastery of religious texts as the height of spiritual attainment. Founded by the charismatic Israel ben Eliezer, known as the *Baal Shem Tov,* the possessor of the secret of the Divine name, the movement spread rapidly. It became especially popular among the poorer, uneducated classes, who learned that a profound love of God expressed through intense, joyful prayer, song, and dance was no less meaningful than the scholarship of the intellectual elite. Hundreds and thousands of devoted Hasidim, or followers, venerated as sacred royalty the saintly disciples of the Baal Shem, men of exceptional piety, righteousness, and wisdom, who in time established spiritual dynasties. The Hasidic leaders came to be regarded as intermediaries and miracle workers with direct access to the heavenly world. Instead of the traditional title of rabbi, suggesting legal expertise and authority, they took the title of Rebbe, the colloquial Yiddish form of the word, suggesting spiritual and moral leadership.

Soon after Menachem Mendel left his Rebbe, he found his place in Pshischa and flourished as the brightest and most diligent of the students there. He drank thirstily from the wisdom of the Hasidic masters and eventually took up the mantle as leader of his own court, attracting followers and disciples to the town of Kotzk. Menachem Mendel, the new Rebbe of Kotzk, barely forty years old, had an intellect sharp as a razor, caustic wit, and a harsh, impatient, demanding manner with his followers. Mordechai Yosef Leiner, from the town of Tomashov, himself already a great rabbi, joined the disciples who followed Menachem Mendel to Kotzk.[89]

From 1826 to 1839 Leiner, called "the lion of the troop," became Menachem Mendel's foremost disciple. But the difference in outlook between him and his teacher gnawed at his conscience and troubled his soul. The Kotzker Rebbe believed that the path to holiness led through scorching introspection and relentless purification of every thought and action. True, not many could practice that kind of rigor and self-analysis, but the quality of those few faithful balanced the quantity of students who could not stand in the glow of the Rebbe's heat. This elitist philosophy chafed Leiner's temperament, and he did not believe that it corresponded to the teachings of the founding masters of Hasidism. After all, the movement itself had railed against the rabbinical elitism that focused exclusively on textual excellence. As the years passed, the philosophical gap between Leiner and Menachem Mendel widened until the pressure became too much for the disciple to bear.

On the eve of the Festival of Rejoicing with the Torah, ecstatic worshippers traditionally danced around the synagogue, lovingly embracing the holy Torah scrolls. Seven such circuits are customary, each headed by an honored leader. Leiner had always led the sixth and most coveted round of dance – but not this year. The change signaled to Leiner that the Rebbe finally had enough of the disciple's contesting the ideology of the sect; his opinions and ideas had taken him beyond the pale.

The dance became a metaphor for the relationship between mentor and disciple. Leiner no longer had a welcome place in the circle. The blatant affront ignited years of pent-up frustration, and Leiner rounded up a group of students and abruptly departed the festivities. He did not return the next day nor the following Sabbath eve. The Kotzker Rebbe was distraught over the departure and secession of his foremost student. It challenged the entire philosophy of Kotzk, for, if the goal was to perfect the select few, surely Leiner represented the elite of the Kotzk disciples. If Leiner could not acknowledge the validity of the Rebbe's ideals, then how could the system as a whole succeed?

Menachem Mendel broke. He approached the Sabbath table – and what happened next has become the subject of passionate controversy.[90] According to one version, in an unthinkable act the pious Rebbe knocked over the candlesticks, violating a most basic principle of Jewish law by extinguishing a fire on the holy Sabbath.

"There is no Law, and there is no Judge!" he exclaimed in a fit of rage.

The room went silent. The Rebbe was silent. Then Menachem Mendel left the table, ran upstairs to his room, and there he stayed, refusing to come out, for the next twenty years until his death.

The disciples of the Kotzker Rebbe never forgave Leiner for his betrayal of their master, and the antagonism survived for generations. It had a profound influence on the lives of future leaders of the Radzyn dynasty, including Leiner's grandson Gershon Henokh Leiner.

Gershon Henokh, born in 1839, was a gifted child. When four years old, playing outside his grandfather's window, he overheard a lecture in which the speaker stated that there are ninety-six letters in a certain biblical passage. The young Gershon

Henokh casually remarked that this statement held true as long as you didn't count words mentioned more than once in the passage. He became a brilliant scholar at a young age and at sixteen began to compile a commentary on the most arcane tractates of Jewish law. So masterful was his work, which took him ten years to complete, that the one criticism leveled against it was that people might confuse it with the holy Talmud.

After his grandfather and father died, Gershon Henokh succeeded to the leadership of the Hasidic dynasty of Radzyn. Where his father had been cautious and reserved as a Rebbe, Gershon Henokh behaved as a maverick, constantly generating new and radical ideas.

"Why can't you be more like your father?" people chided.

To which he answered, "What do you mean? I am precisely like my father. My father was exactly the opposite of his father in temperament, and so my temperament is the opposite of my father's!"

Seal of the Radzyner Rebbe
ARI GREENSPAN. | TEKHELET.COM

Gershon Henokh had an insatiable curiosity and an unfettered intellect. He quickly mastered not only the vast and complex course of study for the Jewish scholar, becoming an expert in many of the more obscure details of the law, but with equal avidity he absorbed a great deal of the secular knowledge of his day as well. A true autodidact, he picked up enough engineering skills to design a flour mill that produced thousands of pounds of flour a day. He also taught himself pharmacology and wrote prescriptions accepted by Polish apothecaries.

In 1887, Gershon Henokh came to a dramatic decision. It was no longer acceptable, he thought, that Jews should continue to ignore a basic biblical precept. The time had come to learn once again how to produce the *tekhelet* dye used in ancient times to make sky-blue tzitzit, which meant identifying the *hillazon*, the mysterious creature from which the dye was extracted.

The circumstances surrounding the loss of the art of *tekhelet* dyeing to the Jews of Palestine in the seventh century had also led to the loss of knowledge about the marine source of the dye. Although Pliny and others had written of the *purpura, conchylium*, and *Buccinum*, with their "grooved and projecting muzzles," and the Talmud depicted the *hillazon* as being "similar to the sea," these general descriptions offered little help. Genuine expertise in identifying the exact animal and certainly any practical information about catching the creatures, obtaining the dye, and carrying out the dye process had faded into distant, obscure, and unreliable memory. Furthermore, although some Jewish authorities had associated Tyrian purple with biblical *tekhelet*, this belief by no means represented the only, or even the predominant, opinion. Many believed that the *hillazon* did not belong to the natural order at all but rather was a singular miraculous creature designed specifically to produce the unique *tekhelet* dye ordained only for use by the Jews. In a wonderful bit of Talmudic sophistry, some rabbis even proved that according to the Bible the *tekhelet* used for the high priest's garments differed from that used for the fringed tzitzit of the layperson, each coming from a different source, both animals termed *hillazon,* both dyes called *tekhelet,* and both the same azure color of the clear sky.

Some Christian Hebraists in the eighteenth and nineteenth centuries associated the *hillazon* with snails, and Wilhelm Gesenius, the great German biblical scholar and semiticist, wrote of a *Purpurschnecke,* a purple-producing snail that yielded a dye of blue-violet color. Rabbi Israel Lipschitz, a leading legal scholar also interested in the problem, knew of Gesenius's idea but rejected it because he believed that the authentic color of *tekhelet* was sky blue, with no hint of purple or violet. Leiner probably knew of this scholarship as well but agreed with Lipschitz and therefore never considered the murex. Faced with uncertainty regarding the nature and identification of the *hillazon,* the Radzyner Rebbe took it upon himself to search for the source of *tekhelet* in order to reinstate the observance of the commandment of wearing blue fringes.

But why this sudden, urgent desire to rediscover knowledge lost and essentially disregarded for centuries? With the early stirrings of Zionism in the mid-1800s, some Jewish thinkers began to wonder what a Jewish homeland would look like and what new opportunities it would afford for religious observance. For the religious Jew, the long-yearned-for rebuilding of the Temple and the reinstitution of the sacrifices and all other priestly tasks represented the climax of the return to Zion. But Temple services couldn't be performed without the sacred garments of the high priest – and the law was very clear; those garments required *tekhelet.*

Rabbi Lipschitz, an older contemporary of Gershon Henokh, had suggested that sky-blue dye from any source could fulfill the requirement for *tekhelet,* provided that the dye held fast and didn't fade or wash out. Leiner disagreed, however, adamantly maintaining that only a dye obtained from the *hillazon* sea creature qualified – thereby creating an apparently insurmountable obstacle to the hopes and dreams of generations of Jews. Overcoming this obstacle may have been the impetus for the desire to identify the *hillazon* as a prelude to rediscovering the methods of *tekhelet* dyeing, a precondition for the reestablishment of God's house and His worship.

And so, Gershon Henokh Leiner, the grand Rebbe of Radzyn, set out to find the marine source of the *tekhelet* dye. Together with his son and an aide, he left his family, his congregation of devout followers, and his home in Poland and traveled south to the Italian port city of Naples.

Just a few years prior, the Aquario de Napoli had been founded, the first aquarium of its kind opened to the general public. There, in the greatest collection of Mediterranean sea life available to scientists or the general public, Leiner began his research in hopes of discovering the identity of the elusive *hillazon*. Filled with wonder and admiration, he wrote of his impressions in the old-fashioned, accumulative, rabbinical style: "And there are rows of new buildings built of clear glass, and the sea water is constantly flowing, and there all the creatures of the sea move about freely, and all is exposed and visible behind the glass walls, and there one can observe every animal, its conduct, eating habits and all the details of its behavior."[91]

Leiner spent many months studying the creatures swimming about in the tanks of the great aquarium, looking to see if the often vague clues dispersed throughout the Talmud, written over

ARI GREENSPAN. | TEKHELET.COM

Ticket to the Naples aquarium, late 1800s

a thousand years earlier, applied to any of them. Some of the Talmud's more obscure and difficult comments contradicted the facts of nature: "The *hillazon* comes up [from the depths of the sea] once every seventy years," for example.[92] Leiner dismissed those as hyperbolic passages not meant literally.

But other Talmudic passages, more descriptive of physical traits, he took seriously. "The *hillazon*, as long as it is alive, its covering grows with it,"[93] which meant that only a fish with a shell or cartilage casing could count as a viable candidate, and "It has curled protrusions or pegs coming out of its head."[94] In all, he enumerated ten characteristics that a *hillazon* categorically must possess, describing them in a brief monograph that he wrote shortly before leaving Poland.

Eventually the grand Rebbe of Radzyn settled upon the cuttlefish, a type of squid, and declared that odd creature the long-sought *hillazon*.

Sepia officinalis has many of the qualities mentioned in the Talmud; most significantly it secretes an ink-like liquid to confound enemies when in danger. Writers had used the pigment derived from the ink sac of the cuttlefish for centuries; in fact, in German the cuttlefish is called *der Tintenfisch* – the ink fish. In the world of art, the sepia tincture colored the works of the greatest European painters, including Leonardo da Vinci and Rembrandt van Rijn.

Convinced that he had found the long-lost *hillazon,* Leiner faced one major and potentially insuperable problem: color. Sepia is a shade of brown, but *tekhelet* was supposed to be sky blue. Not to be put off by this impediment, the great Rebbe presumably consulted with chemists who apparently offered him advice and suggestions. After some effort and experimentation, he actually managed to produce a blue dye from the cuttlefish ink.

Leiner had succeeded in revealing a long-buried truth. His mission complete, he returned to Radzyn triumphant. In the early winter of 1891, he set up a dye factory that began producing blue threads for the tzitzit. Within a year, he had manufactured more than ten thousand sets, and his faithful followers wore them prominently.

Not everyone agreed with Leiner, however. Attacks on his identification of the *hillazon* came almost immediately from various sides. Some of the attacks had political overtones. The Hasidim of Kotzk still hadn't forgotten or forgiven how Leiner's grandfather had abandoned their master in order to establish his own dynasty. They led a vicious assault against the new *tekhelet,* often more personal than pertinent.

But others leveled more concrete and germane arguments against Leiner's assertion. In particular, two key problems troubled opponents and seemed to disprove the cuttlefish identification. First, the blue dye from Radzyn faded over time, and soap could wash it out, whereas tradition maintained that true *tekhelet* was dye-fast and "retains its beauty," as the great legalist Maimonides put it. [95] Moreover, the Talmud had emphasized that dye from the *hillazon* had to be obtained while the creature was still living – "the more life it has, the better the dye" – whereas Leiner used dead cuttlefish in his process. [96]

To defend his position and counter the criticism mounting against him, as well as to provide more detailed information on all aspects of the laws of *tekhelet,* Leiner wrote several books on the subject, the last of which appeared posthumously. The great Rebbe of Radzyn died not long after his magnificent discovery, at the young age of fifty-two. He had lived long enough to fulfill his deepest desire and saw tens of thousands of Jews wearing the blue fringes that validated his efforts.

His mantle of leadership and the responsibility for defending Radzyn *tekhelet* passed to his son, Mordechai Yosef Elazar Leiner, who had accompanied his father on that fateful trip to Naples and who prepared the Rebbe's last and most comprehensive work on the topic for publication. When he died in 1929, his son, Shmuel Shlomo, took over, the last male-line descendant of Mordechai Yosef Leiner, the founder of the dynasty.

In the spring of 1941, as the Jews of Radzyn were herded into ghettos, the hope of survival was fading. Shmuel Shlomo Leiner, known affectionately as Reb Shloimele Radzyner, realized that the correct course of action for his disciples was to escape to the vast Polish forests and join the partisans in their resistance against the Nazis. Word of his decision quickly reached the authorities. When they began to look for the Rebbe, his closest confidants counseled him to leave Radzyn and hide in nearby Vlodova.

The gestapo spent weeks searching for him, to no avail. One morning, the Jews of the Radzyn ghetto found themselves surrounded by troops who randomly opened fire, killing dozens. The commander sent word to the Jewish leadership that unless they immediately handed over the Rebbe of Radzyn, all the Jews in town would be executed. Without hesitation, Yaakov Wolf, Reb Shloimele's faithful aide, donned the Rebbe's coat and surrendered to the Nazi officer, claiming that he was Reb Shloimele. The commander promptly shot him dead, and the soldiers withdrew.

Within a few days, however, the gestapo became aware of the ruse and learned that the Rebbe had fled to Vlodova. Troops shortly surrounded the ghetto of Vlodova with the same demand: Give up the Rebbe or die. Reb Shloimele, who had not learned of the events in Radzyn, wrapped himself in the prayer shawl passed down from his grandfather Gershon Henokh and walked out to the Nazi line. They marched him to the Vlodova cemetery, and as he reached the gate, an officer roughly pushed him through it. Reb Shloimele Radzyner turned around and spit in the Nazi officer's face. With burning rage, the Nazi shot him dead on the spot. Reb Shloimele fell to the ground, his crimson blood staining the first set of *tekhelet* strings that his grandfather had dyed in Italy some fifty years before.[97]

The controversy over the Radzyn *tekhelet* raged on long after Gershon Henokh's time. The opponents' resolute disapproval often turned aggressively antagonistic. People asked prominent rabbis and scholars whether adherents who wore the new *tekhelet*

could be permitted burial in the same cemeteries as other Jews who kept the tradition of exclusively white tzitzit.[98] The controversy divided families and caused discord and acrimony within communities. It even reached beyond Poland all the way to the Old City of Jerusalem.

There, one pious and ascetic rabbi, Hillel Gelbshtein, who long before had studied in the court of Menachem Mendel of Kotzk for a few months before the Rebbe's death, led a bitter campaign against the new blue. He harshly called for the removal from public office of anyone who wore it and for disqualifying them from offering testimony in court. In one particularly zealous moment, Gelbshtein went so far as to tear the prayer shawl off an eminent individual who wore the Radzyn *tekhelet* in public. Then he set it ablaze in the courtyard of the synagogue.

Opponents continued to criticize, but the outpouring of hostility didn't deter the Radzyn Hasidim. Yet the essential questions ultimately remained in dispute: Was the blue dye that Gershon Henokh had discovered and manufactured authentic *tekhelet*, and was *Sepia officinalis* the true *hillazon*?

It fell to another brilliant, eclectic, self-educated scholar and rabbi to try to resolve the issue once and for all.

IX
BLUE BLOOD

*S*lonim, a small town in Belarus, had a large Jewish population headed by a rabbi universally acknowledged and respected as an expert in the intricacies of Talmudic scholarship and an authority on Jewish law. As such, he often received letters asking for his opinion on a wide variety of religious matters. But one letter baffled and intrigued him. It did not contain the usual request for legal judgment; rather it was an erudite commentary on aspects of the rabbi's own work. Even more unusual, the correspondent – unmistakably brilliant and thoroughly conversant with the whole range of Jewish law and custom – had sent it anonymously.

The letter came from Leeds, in the north of England, far removed both geographically and intellectually from the famous eastern European centers of Jewish life and learning. But the text was written in the traditional rabbinical Hebrew, a kind of lingua franca that over the centuries had enabled Jews the world over to communicate with each other about religious and other issues. The rabbi of Slonim knew of all the leading scholars in his day, but he knew no one in Leeds who could have displayed the broad scope and profound insights contained in the letter. When he eventually managed to discover the identity of his correspondent, he was amazed to find a young man who was not yet out of his teens, who had never studied formally in a traditional yeshiva, and whose modesty prevented him from mentioning his name. Born in Poland in 1888, Isaac Halevi Herzog – his middle name testifying to descent from the biblical tribe of Levi – moved to England with his family when he was nine. His father became a rabbi in Leeds and personally undertook Isaac's intensive religious education rather than sending him to school. A prominent Jewish scholar traveling in England later subjected young Isaac to rigorous examinations and found him to be one of the greatest Talmudists of the times. Before turning twenty, Herzog received official rabbinical ordination from three leading Talmudic scholars, including the rabbi of Slonim.[99]

A true prodigy, the young Herzog didn't fit the mold of scholarly genius typical of the narrower world of traditional eastern European Jewry, who tended to fear and avoid modernity and its secular knowledge. Herzog's education did not cease with his mastery of Talmudic knowledge. He studied at the Sorbonne in Paris and earned a degree in Oriental languages. He earned degrees in classics and mathematics at the University of London, where in 1914, at the age of twenty-five, he also received a doctorate. Investigating what he called porphyrology, his thesis was "The Dyeing of Purple in Ancient Israel."

Herzog took up the position of rabbi in Belfast in 1916, the year of the Easter Rising, and three years after that he moved to Dublin, eventually becoming chief rabbi of Ireland, a position he held until 1937. During those decades, Ireland experienced many upheavals – the Irish War of Independence, civil war, and the creation of the Irish Free State. Herzog viewed the Irish nationalist movement for self-governance and revival of the ancient

Gaelic language as similar to what the Zionist movement pre-scribed for the Jewish people. His sympathies lay with the nationalists rebelling against British authority, and he developed a close friendship with Éamon de Valera, the political leader of Sinn Féin and later prime minister and president of Ireland.[100]

De Valera urged him to learn Gaelic, in which Herzog be-came fluent. His years in the fledgling Irish Republic during that troubled period deeply affected his attitudes on social issues. A lifelong champion of the underdog and the underprivileged, he wrote editorials in the Irish press calling for social programs to alleviate the plight of the poverty-stricken residents of Dublin.[101]

During his years in Dublin, Rabbi Herzog continued his scholarship, writing research papers on a wide array of topics from law to philosophy and, of course, on Jewish thought. When Rabbi Abraham Kook, the revered chief rabbi of the British Man-date for Palestine, died in 1935, Rabbi Herzog's name came up for consideration as his successor. Though extremists opposed the appointment of a chief rabbi so dangerously familiar with the world of secular knowledge – posters in the Holy City proclaimed JERUSALEM WILL NOT ABIDE A RABBI-DOCTOR AS ITS HEAD – most rabbinical authorities heartily endorsed his candidacy.[102] Herzog won the election and took the position (which he held from 1937 until 1948, when the UN partitioned the British Man-date and created Israel, after which he served as the chief rabbi of Israel until his death in 1959).

With the outbreak of World War II, Herzog focused his ef-forts and influence on saving the Jews of Europe. He vehemently opposed the British White Paper that limited the rights of Jews to emigrate to Palestine, claiming that it would seal the fate of tens of thousands who would have no safe haven from the Nazis.

In 1941 he traveled to America and met with President Roosevelt to plead for more direct intervention against the slaughter of Jews. Though Herzog left the meeting disappointed, he had made a great personal impression on Roosevelt, who later called the rabbi to try to convince him that it was too dangerous for him to return to Palestine. Field Marshal Rommel was advancing across North Africa, and the possibility of his crossing into Egypt and then to the Holy Land looked both real and immediate. Herzog, however, a man of strong faith and conviction, responded that, while he appreciated the president's concern, there was no reason to worry. Palestine, he firmly believed, was under divine protection, and Rommel's army would not reach it.[103]

During the long and arduous journey back to Palestine, he and his wife narrowly escaped the torpedoes of German U-boats, then booked passage on the last civilian ship to cross the Atlantic during the war. They made their way through Africa from Johannesburg to Mozambique, Khartoum, Cairo, Alexandria, and across the Suez Canal in an adventure worthy of a Hollywood script. After the war, Herzog traveled throughout Europe searching for and recovering surviving Jewish orphans from various hiding places, enlisting the help of political and religious leaders, including Pope Pius XII, Queen Wilhelmina of the Netherlands, and the prime ministers of Belgium, France, Ireland, Italy, and Poland.

A tireless advocate for the establishment of Israel, Herzog in 1949 met with Harry Truman, the first world leader to acknowledge the sovereignty of the new nation, to thank the president for his support. David McCullough, in his book *Truman,* describes the meeting.

When the chief rabbi of Israel, Isaac Halevi Herzog, called at the White House, he told Truman, "God put you in your mother's womb so that you could be the instrument to bring the rebirth of Israel after two thousand years." David Niles, political advisor to Truman, recalled the meeting: "I thought he [Herzog] was overdoing things… but when I looked over at the president, tears were running down his cheeks."[104]

Statesman, rabbinic sage, ardent defender of the rights of the underprivileged: Herzog was all of these. But throughout his life, since his days as a student in London, one subject held a very dear place in his heart and mind, the theme of his doctorate. Hebrew porphyrology – the term Herzog coined for the study of the ancient biblical dye *tekhelet* – remained his enduring passion. The dyers of Radzyn had in fact contacted his father, Joel, sending the venerable rabbi of Leeds details regarding their dyeing process in the hope that he would give the endeavor his approval and support.

Herzog's doctoral dissertation displays a mastery of such diverse subjects as archaeology, Greek and Roman literature, chemistry, Talmudic and Midrashic texts, and philology, and it includes references to Semitic languages, Sanskrit, and Chinese. According to his son Chaim Herzog the rabbi had a good knowledge of some twelve languages. When U Nu, the first prime minister of Burma (now Myanmar), visited Israel, the only official to greet him in his own language was Herzog. The bulk of his dissertation relates to assessing the possibility that the long-sought *hillazon* was indeed a sea snail. One fascinating portion of his dissertation, however, relates to the work of Gershon Henokh

Rabbi Herzog returning to Israel after meeting with President Truman, 1949

Leiner, the Radzyner Rebbe who had identified the enigmatic *hil-lazon* as the cuttlefish *Sepia officinalis.*

Rabbi Herzog's research and his reading of the classical sources dealing with ancient dyeing practices argued strongly against the Radzyner's theory, but Herzog sought empirical evidence before coming to a definitive conclusion. He obtained samples of the Radzyn *tekhelet* strings and sent them to German chemist Paul Friedländer for analysis. Known for his research on indigo, Friedländer in 1909 had extracted pure Tyrian purple from the *Murex brandaris* and described the structure of the dye molecule. There was no better man for the job.

The results of Friedländer's analysis of the Radzyn dye were hard to believe:

Because of a lengthy illness I was unfortunately prevented from answering your very interesting letter earlier. As far as the dyeing process from Radzyn is concerned, what is lacking there is the main thing, namely the dye stuff with which the wool thread is coloured. That is a perfectly ordinary modern tar based dye. The details about the use of Sepia officinalis *are only misleading, and it is completely impossible with this substance to achieve that kind of colouring. In addition, I also consider it impossible to produce a pure blue from the purple snails that are known to me.*[105]

The letter astounded Herzog. Friedländer was stating unequivocally as his professional judgment that the blue dye that Leiner had so painstakingly prepared from cuttlefish ink could not have come from that creature. The additional comment about the impossibility of getting a blue dye from the murex snails also greatly influenced Herzog's ultimate conclusions.

Seeking corroboration for Friedländer's radical assertions, Herzog sent strings for additional tests to the Manufacture des Gobelins, the French factory that produced the famous Gobelin tapestries. The results were essentially the same:

According to your wishes I have examined the sample of blue that you sent me. I don't know of any natural blue colour other than indigo that is capable of solidly dyeing textile fibers. As far as the Sepia is concerned, it yields brown and not blue, and that brown furthermore is not suitable for fixing on fiber. The blue of the sample that you sent me exhibits all of the characteristics of Prussian blue obtained by processing salts of iron and potassium.[106]

These two independent sets of tests proved conclusively that the Radzyn dye was inorganic. It was, in fact, as the French test showed, a synthetic dye known as Prussian blue. Although his own studies had indicated that the *hillazon* was most probably a type of shellfish, the situation utterly perplexed Herzog. Surely, the great Rebbe of Radzyn, engaged in holy work, would never intentionally have created a spurious product to mislead his followers. Determined to uncover the facts of the matter, Rabbi Herzog sent away to the dye masters of Radzyn for the exact recipe used in producing the dye.

The Radzyner Rebbe himself had died over twenty years earlier, but his followers dutifully maintained the tradition of dyeing *tekhelet* and wearing it on their tzitzit. The correspondence that Herzog received began with the firm declaration that the *tekhelet* dye color came exclusively from the cuttlefish ink since all other ingredients added during the process were either colorless, gray, or white. "Only the blood of the *hillazon* contained in its bladder is black like ink." This "blood" they placed in a heavy cauldron to which they added grayish iron filings together with a chemical "which is white as snow called potash." They heated the mixture for five or six hours on a very high flame "until the fire burns inside and out like the fires of Hell," fusing the blood, iron, and other chemicals together.[107] The final product was the blue dye of Radzyn.

The disciples of the Radzyner Rebbe knew how to follow his recipe to the letter – but they had no knowledge of chemistry or of the nature of the processes that occurred in their cauldron. Notwithstanding their assertions, the blue color they obtained had nothing to do with the ink from the cuttlefish and everything to do with the iron filings. Unaware that they were doing so, the Hasidic dyers were actually manufacturing Prussian blue.[108] Berlin Blue, as it was originally called, was the first synthetic blue pigment. In 1704 a German painter and color maker, Heinrich Diesbach, was trying to create a cochineal red dye-based pigment. While working in the laboratory of an alchemist, Johann Dippel, he accidentally used some contaminated potash. Instead of the expected red, to his great surprise and ultimate delight, he ended up with a marvelous, deep blue.

The contaminant responsible for this fortuitous outcome was something called Dippel's Animal Oil, a malodorous potion that the alchemist hawked as an elixir of life. But Dippel was no ordinary snake-oil salesman. A theologian, prolific author of combative religious tracts, successful physician, and a practicing alchemist, he also gained a reputation for dissecting animals and experimenting with cadavers to explore the possibility of transferring souls. He concocted his wildly popular animal oil from blood, flesh, bones, and an assortment of chemicals, offering an improved version of this elixir of life to the owner of Castle Frankenstein near Darmstadt in Germany in return for the castle itself – an offer that the castle owner wisely declined. Dippel had been born at the castle, and the coincidence of the castle's name, his experiments, and a trip by Mary Godwin (later Shelley) to the nearby town of Gernsheim have made it inevitable that speculation should arise concerning the novelist's knowledge of Dippel and his life. Neither the rumors about Dippel nor the speculation about the literary genesis of Shelley's novel *Frankenstein* stand on very solid ground, but Diesbach's discovery of Prussian blue has had a profound influence on the visual arts.

Painters wanting blue for their work had few convenient or affordable choices. Indigo on canvas didn't retain its color well, and ultramarine, derived from the semiprecious stone lapis lazuli, was inordinately expensive. Prussian blue, with its mix of carbon, nitrogen, and iron atoms, is known chemically as ferric ferrocyanide and can be manufactured easily and in large quantities. The ingredient typically added to provide carbon and nitrogen was ox blood. After its discovery, painters almost immediately took advantage of the new color. In 1709 Pieter van der Werff used it for his *Entombment of Christ,* and, a century later and halfway across the world, Japanese artist Katsushika Hokusai used it to spectacular effect in *The Great Wave off Kanagawa.*[109]

Although mainly important as a dye or paint, Prussian blue has some curious and unexpected uses as well. According to its website, the Centers for Disease Control and Prevention (CDC) has included Prussian blue in the Strategic National Stockpile, which is "a special collection of drugs and medical supplies that CDC keeps to treat people in an emergency." Prussian blue prevents the intestines from absorbing radioactive cesium and thallium into the body, so that the harmful elements can pass through instead (resulting, perhaps not surprisingly, in blue feces).

Rabbi Herzog's realization that Radzyn *tekhelet* was in fact synthetic Prussian blue led him to reject it categorically as the authentic biblical dye and to deny the identity of the cuttlefish as the elusive *hillazon*. It was inconceivable, argued Herzog, that the Talmud so adamantly insisted on the *hillazon* as the sole permissible source for *tekhelet* if the marine animal played only a nonessential role in the dye formation. Why require cuttlefish ink rather than ox blood if both – or countless other materials, for that matter – could generate the same result?

In a twist of historical irony, with the extermination of Eastern European Jewry during World War II and the destruction of the *tekhelet* factory also went the knowledge of the controversial Radzyn dyeing process. When survivors of Radzyn made their way to Israel after the war, they asked Rabbi Herzog for the correspondence between his father and the Radzyn dye makers. The detailed information in those letters enabled them to reestablish a Radzyn *tekhelet* industry in Israel, which still exists to this day. As a result, Herzog definitively discredited Radzyn's *tekhelet* while simultaneously rescuing the process from oblivion.[110]

In his dissertation, Herzog amasses evidence pointing to the *Murex trunculus* snail as the most likely candidate for the long-sought *hillazon*. He knew of the numerous archaeological expeditions during the nineteenth century up and down the Mediterranean coast from Jaffa to Tyre that unearthed piles of crushed murex at various sites. He knew that archaeologists there had always uncovered two distinct mounds of shells: one of *brandaris* and *haemastoma* together, and the other of *trunculus*. He read Aristotle and Pliny (in the original Greek and Latin) and recognized their descriptions of the purple-producing snails as the murex. He had familiarized himself, too, with the work of Lacaze-Duthiers and knew full well that the scientific community had declared the case closed and the puzzle of ancient shellfish dyeing solved. Tyrian purple came from *Murex brandaris* and *Thais haemastoma* (a reddish purple), and *tekhelet* came from the *Murex trunculus* (a blue purple).

But Rabbi Herzog had reservations about the conventional wisdom and hesitated to offer an unqualified endorsement of the *trunculus* as the authentic *hillazon*. First and foremost, he was trying to solve the riddle from a religio-legal perspective rather than a purely historical or scientific one, so he had to work within the constraints of that system, its sources and precedents found primarily in the Talmud. But those passages relating to the *hillazon* are ambiguous and prone to hyperbole. For example, the rabbi's first problem with identifying the *hillazon* with the *trunculus* snail was the Talmudic statement that the *hillazon* is "similar to the sea." What exactly does that mean? What attribute of the creature does this similarity indicate? Herzog did not consider this characteristic as absolutely critical, but he raised it as an issue nonetheless since the only specimens of murex that he had seen had been cleaned and polished, and did not in any way resemble the sea. When in their natural habitat, however, the snails develop a covering of small creatures and plants, plus a greenish coating, that renders them identical to the rocks and other surroundings of the seabed.

Furthermore, the *trunculus* does not fit the Talmud's description of a creature "coming up once in seventy years." No animal known to science fits such a description.[111] Nor did chemists at the time believe the murex extract to be a steadfast, long-lasting dye. Subsequent research has disproved this assertion, establishing that murex dye is among the fastest dyes in nature and certainly the strongest and most lasting known to the ancient world.

The primary argument against the *trunculus,* however, was color. *Tekhelet* had to be sky blue. This was a long-held Jewish tradition that Rabbi Herzog could not dismiss – but the dye obtained from *Murex trunculus* appeared bluish purple. All of the great scholars and scientists, from Gesenius, who wrote about it in his Hebrew Lexicon in the early 1800s, to Friedländer, who discovered the purple molecule, agreed on that point. Herzog thus concluded that, "if we unequivocally determine that the appearance of *tekhelet* had no violet [purple] component, then this would be enough to dislodge the assertion" that *Murex trunculus* is the *hillazon.*[112]

Not that everything in the Talmud pointed against the *trunculus.* For instance, the Talmud indicated that the *hillazon* could be found "from the Ladders of Tyre to Haifa," clearly meaning the Mediterranean.[113] "It has a covering that grows along with it" implies that it was indeed a shellfish.[114] That certain Talmudic statements about the *hillazon* conflicted with others, some corroborating and some challenging an identification with the *trunculus,* led to Herzog's ambivalence. He entertained the idea that perhaps the *hillazon* might be some other kind of sea snail. A beautiful and delicate mollusk of the *Janthina* genus was one candidate, but no archaeologist has ever found evidence that they were collected in antiquity, and no chemist has ever succeeded in producing a dye from them. That suggestion was really nothing more than hopeful speculation, and the problem of identifying the *hillazon* haunted Herzog for the rest of his life.

The dream of this modern, intellectually sophisticated, utterly devout rabbi resembled that of the saintly Hasidic master Gershon Henokh Leiner before him: to restore the possibility of fulfilling the ancient commandment of wearing *tekhelet*. As a perceptive secular scholar, Herzog watched the evidence increase in favor of the *trunculus*. Not bothered by the dye's purple tint, scientists simply defined the color of authentic *tekhelet* as violet. But Rabbi Herzog remained loyal to tradition as well as science, and those conflicting loyalties held him back from accepting the unanimous verdict of the historians, chemists, and archaeologists. Lest one think that this verdict was based on the immature science of the early twentieth century, as late as 1979, the greatest naturalists were still adamant in their assertion that sea-snails produce only purple. In response to a request for current information about possible marine sources that might yield a blue dye, a scientist at the world renowned Observatoire Océanologique de Banyuls-sur-Mer wrote, "I sent your letter to a tissue dye specialist at the Museum in Basel, Switzerland, hoping he would know the origin of the dye at once. But nor him nor his colleagues have the faintest idea. They only know of 2 or 3 dyes coming from marine animals, and these are all purple."[115]

Rabbi Herzog never could reconcile the dissonance between the secular and the traditional positions and remained ambivalent about the identity of the *hillazon*. A reconciliation would come, but not for nearly a quarter century after his death in 1959.

Like so many important scientific developments, particularly involving dyes, serendipity solved the vexing problem of the *trunculus* dye and the traditional *tekhelet* color.

The story of the reconciliation begins with a boy genius from Chattanooga, Tennessee. When experiments with homemade fireworks quite literally blew up in his face and sent him to the hospital, Sidney Edelstein's nascent interest in chemistry nearly ended prematurely. Undeterred, however, he entered MIT at the age of sixteen and became an expert in the chemistry of textiles. On graduating, at the height of the Great Depression, he returned to Chattanooga looking for a job and worked to bring the science he had learned to the unsophisticated factories of the Volunteer State. At night he continued his research, expanding his knowledge of textile fibers to include the practical chemistry of dyes. During World War II, he invented Kopan, a material that considerably improved the standard mosquito and camouflage netting used by Allied forces.

From being a research chemist and inventor, Edelstein went on to become a major industrialist with the founding of the Dexter Chemical Corporation. His varied interests and his fascination with the history of science and technology eventually led him to book and manuscript collecting. He contributed to the publication of scholarly editions of important works, including the great Renaissance book on dyeing, the first printed book on the subject, by Gioanventura Rosetti, *Plichto de larte de tentori,* published in Venice in 1548. Edelstein himself collaborated on an annotated edition of that work, *Instruction in the Art of the Dyers,* published in 1969, and later donated his vast collection of books and manuscripts to the Jewish National and University Library in Jerusalem.[116]

THE RAREST BLUE

As we saw earlier, in the 1960s, the famous Israeli archaeologist Yigael Yadin had turned to Edelstein for help in analyzing some of the textiles found in the digs at Masada, including the bundle of purple wool, which the archaeologist speculated might be authentic *tekhelet*. It turned out not to be the case, but the find intrigued Edelstein nevertheless. He developed a lasting interest in the ancient purple shellfish dyes and particularly in the Jewish *tekhelet*. On a visit to Israel that had decisive consequences, he asked two professors for help in researching the topic, Ehud Spanier, a marine biologist from Haifa University, and Otto Elsner, a dye chemist from the Shenkar College of Fashion and Textile Technology. Elsner was particularly well suited for the task since as a young chemist he had worked with woad dyeing in his native Poland before immigrating to Israel.

Because of the unavoidable rancid smell that accompanies dyeing with snails, Elsner conducted his experiments near an open window. That seemingly insignificant choice of location led to a groundbreaking observation. Elsner gathered the snails, extracted the glands, and collected the dark purple dye. He applied the standard techniques used in woad dyeing, adding the necessary chemicals to put the dye into solution. He immersed the wool, and after removing it he watched it turn from greenish yellow to its final, permanent color. As expected – and as all scholars and chemists had claimed for a hundred years – the wool indeed took on a purple hue.

On cloudy days, that is.

But when working in bright sunshine, Elsner saw something quite different: The *trunculus* dye produced a beautiful sky-blue color.[117] Furthermore, the azure color of the wool that emerged from the brackish liquid was fast and lasting; it didn't fade, nor did it change color.

Elsner researched the chemical processes taking place in his sun-drenched dye mixture and discovered that while in solution the dye molecules are less stable than when in their natural condition. Ultraviolet energy – such as found in the rays of bright sunlight – can break the bonds that attach some of the atoms to each other within the dye molecule. These photochemical processes alter the purple and red constituents of the murex extract and leave primarily indigo, which is, of course, pure sky blue.

Otto Elsner had solved the riddle.

Ancient dyers, doing their work out-of-doors in the bright Mediterranean sun, surely discovered the role that daylight played in achieving the final color of the dye.[118] *Argamman,* Tyrian purple, came from *Murex brandaris* and *Thais haemastoma,* which yield a deep reddish-purple dye mixture. *Murex trunculus,* on the other hand, containing a large amount of indigo to start, and further purified by exposure to sunlight, produced a different shade. *Tekhelet,* as tradition had maintained, exactly mirrored the color of the sky.

In the years since Rabbi Herzog drew his line in the sand regarding the color of the ancient dye, other evidence for the sky-blue color of *tekhelet* has come to light. We can deduce the color, for example, from the linguistic history of the word. In ancient Mesopotamia, the Sumerian term for blue was *za.gin.na,* which meant "the color of lapis lazuli." The Akkadian translation of the Sumerian term for blue was *uqnâtu,* used extensively to describe the sky. *Uqnâtu* itself is a near synonym of the Akkadian *takiltu,* which is equivalent to the Hebrew *tekhelet.* That development

Cuneiform for blue, pronounced
uqnâtu *or* takiltu
BARUCH STERMAN | TEKHELET.COM

points toward a color equivalency among lapis lazuli, the sky, and *tekhelet.*

Furthermore, the 2,500-year-old Pazyryk saddlecloth unearthed by Sergei Rudenko has a magnificent pattern comprising separate design elements of purple and sky-blue wool, both shown to have come from murex-based dyes. This artifact proves that ancient technology could produce those distinct shades, and that the artisans of old saw them as discrete hues, each a valuable commodity in its own right.

For 150 years, Jewish scholars researching the topic upheld the traditional position that *tekhelet* was the color of the sky against the unanimous opinions of archaeologists, biochemists, and experts in Greek and Roman history. Otto Elsner, however, demonstrated that a simple technique completely within the scope of ancient capabilities could turn the snail dye from purple to azure. What one man discovered after a year or two wouldn't have escaped the curious and sensitive eyes of dyers who practiced their trade over thousands of years.

Rabbi Herzog's faith had inspired chemists and biologists to search more deeply for a thorough understanding of the murex snails and the dyes they produce. That understanding requires some knowledge of snail biology, dye chemistry, and color physics. In a rare and wonderful convergence, science and religion eventually came to the same conclusion.

X
THE SELFISH SHELLFISH

𝒥n the rocky recesses under the waves, the predator lurks. His keen senses poised, he is watching, waiting. Eons of adaptation have honed his ability to identify his prey, to zero in on its exact location, and to sneak up on it silently, undetectably. The unsuspecting victim is lured into a false sense of security, thinking that its camouflage or other defenses will save it. But the predator has capabilities beyond the prey's comprehension, and there is no question as to the outcome.

Then – he strikes.

Clutching his prey, the predator drips poisonous acid onto its hard, protective armor, softening it, dissolving it. With razor sharp, rasping teeth, he mercilessly gnaws away at the weakened protective coat until he finally reaches the soft, vulnerable flesh beneath. Defenses compromised, the unlucky victim has no choice but to yield to a prolonged and painful death as the mighty victor devours him alive slowly… so very slowly… literally at a snail's pace. It can take the tiny murex, a voracious carnivore, up to sixty hours to bore through its victim's shell – quite a long time to prepare dinner.

THE RAREST BLUE

Mollusks (phylum Mollusca) comprise nearly a quarter of all known marine organisms and include freshwater and terrestrial creatures as well. Aristotle in his *History of Animals* had originally called them *malakia* (μαλάκια), "the soft things," and the English word *mollusk* ultimately derives from *mollis*, Latin for "soft." Mollusks fall into three classes: cephalopods (from the Greek for "head-feet"), which include octopus, squid, and cuttlefish; bivalves, including clams, oysters, and mussels; and gastropods (from the Greek for "stomach-feet"), including snails and slugs, because these animals appear, at least to humans, to crawl on their bellies. Some eighty thousand species make up the gastropod class, by far the most diverse and numerous of the mollusks, second only to insects in the total number of species included in a class.

Of the gastropods, the Muricidae family, commonly known as murex or rock snails, number at least a thousand species. They live in oceans around the world, and their characteristically ornate and intricate spiny shell structures make them a favorite of collectors. Like connoisseurs of stamps and coins, conchologists who collect shells are a special breed. To acquire a prized specimen, they will pay what many consider outrageous prices. Though tourists in any local souvenir shop can pick up some very common shells for pennies, a perfect specimen of the extremely rare *Chimaeria incomparabilis* can fetch as much as $20,000.

In my own collection of murex shells, some specimens reach twelve inches long, while a tiny coral-colored one measures just an eighth of an inch. But each shell contains its own world of beauty and fascination. *Chicoreus torrefactus*, the "firebrand" murex, stands out for its bright orange color. The "leafy hornmouth" murex, *Ceratostoma foliatum,* has thin and delicate spines fused together around the shell opening like a paper-thin collar. *Homalocantha anatomica pele,* with its spikes branching out, looks like some fantastical undersea moose with crazy antlers. Perhaps the most striking of the group, *Murex pecten* has many long, white, pointed spines. Aptly nicknamed the Venus comb, it seems perfectly suited for a mermaid to use to comb her long hair.

As you would expect, I have many specimens of *Murex trunculus* in all sizes from different areas around the Mediterranean.

The authoritative World Register of Marine Species (WoRMS) lists this snail, commonly known as the "banded dye murex," as *Hexaplex (trunculariopsis) trunculus.* When Swedish zoologist Carl Linnaeus, the father of modern taxonomy, named the species in 1758 he must have had in mind their rather diminutive "truncated" spines as compared with the magnificent outcroppings of other murexes. In an interesting parallel, in colloquial Hebrew, *trunculus* and its cousin *brandaris* are both called "*argamon*" or purple shells, but the qualifiers "blunt spined" for the former and "sharp spined" for the latter distinguish the two.[119] *Hexaplex* may refer to the six whorls of an adult shell – not including the top spire.

The word *murex* derives from the Greek *myax* (μύαξ) for sea mussel, and Aristotle refers to these snails as murex, making it one of the oldest given shell names. In Latin *murex* came to mean purple fish. Indeed, one of the key characteristics of the Muricidae family is that they possess an intriguing gland that can produce the beautiful purple dye of antiquity. Another distinguishing feature of these snails is their radulae, the scraping, rasping, tooth-like structures that, as meat eaters, they use to penetrate the protective shells of their prey. Classification and subdivision of the murex into genera and species – though not completely agreed upon by taxonomists and subject therefore to frequent change – often comes down to the shape and size of these radulae.

The murex feast on other snails, bivalves, or shellfish, and they have developed a complex method of drilling through the shells of their victims using those sharp radulae. They also possess another nifty gadget, which, lacking a more impressive term, scientists simply call ABO (accessory boring organ). This small gland is usually found in the snail's foot. About one to two millimeters in size, it is normally retracted, but it can be extended as need arises. When the murex wants to eat, it mounts its prey and maneuvers the ABO into a position adjacent to the selected drilling spot. The foot forms a watertight seal against the target shell around the ABO, which begins to secrete a cocktail of hydrochloric acid, along with enzymes that increase the acidity of the trapped seawater. This mixture softens and dissolves the calcium carbonate shell, at which point the snail shifts to mechanical means of boring. It moves its proboscis – a long trunk-like tube with the radula at the end – into place and scrapes away the weakened layer of shell, which it then ingests. It repeats the process, alternately using the ABO to soften the shell and then the radula to remove layer after layer.* Typical boring rates measure between a quarter to half a millimeter a day, about the thickness of

---

* Here's a fun experiment you can try from Fun with Malacology in Your Kitchen, a hypothetical book. Try to etch or scrape a regular seashell using a blunt knife or sandpaper. It shouldn't even leave a mark. Soak that same shell in vinegar for two or three days. Then try to scrape it again. This time, the shell should be soft, and you should be able to scrape and etch it deeply. Vinegar has roughly

a paper plate, so it can take a snail as much as three days to get to the main course.[120]

Under normal conditions, when food is abundant, murex prefer to dine on smaller prey with thinner shells. But when nothing else presents itself, these snails resort to cannibalism. This gruesome habit has helped archaeologists piece together the process of snail collection for dyeing in ancient times. Examination of discarded murex shells in former dye sites such as Tyre and Dor often reveals telltale bore holes, about two millimeters in diameter, indicating that a fellow murex attacked the shell. On the basis of that evidence archaeologists surmise that dyers caught the snails in large batches and put them into underwater pens to keep them alive until they were ready to be used. The large rectangular pits just under the water level at Dor presumably functioned as such holding pens. Confined and unable to hunt for their usual food sources, the snails turned on each other, leaving small holes in their shells as proof of their captivity two thousand years ago.

The meat that the murex slowly consume, whether of a fellow murex or some other sea creature, they then digest, presumably also slowly. A long process that starts with digestion and continues through a series of complex biochemical reactions results in the blue and purple dyes ultimately obtained from the snails. One byproduct of digestion is a molecule called indole, a

---

the same pH – a measure of the level of acidity – as the liquid created by the murex's ABO.

breakdown product of the essential amino acid tryptophan.[121] The body cannot generate tryptophan by itself and therefore must ingest it. Found abundantly in red meat and seafood, tryptophan helps form the neurotransmitter serotonin, which contributes to the feeling of well-being and happiness, and melatonin, the sleep hormone. This may explain why some snails seem so laid-back and content.

In humans and other animals, excreted feces contain high concentrations of indole, the main contributor to the characteristic odor. Paradoxically, however, in very low concentrations, it has a pleasant jasmine scent and has become a major ingredient in many perfumes.

While human beings excrete indole, the murex snail turns it into a molecule called tyrindoxyl sulphate, a crucial stage in the ultimate creation of the dye. This transformation involves adding bromine to the indole. Scientists wonder about the snail's use of bromine for this process because chlorine has essentially identical chemical properties and is much more readily available in seawater. Murex even have a specific enzyme, bromoperoxidase, that helps them extract bromine from their environment. Whatever advantage the snail may derive from using that element, mankind has benefitted from it because it is the bromine that helps create our lustrous dyes.

Bromine has many other uses as well. From the Greek *bromos* (βρόμος), meaning "stench," bromine indeed lives up to its name; its bleach-like smell can cause eye irritation and coughing even at low concentrations. Since bromine suppresses certain combustion reactions, organic bromine compounds play a major role as fire retardants (known as BFRs, brominated fire retardants). BFRs reduce flammability in plastics, printed circuit boards, and even clothes. Bromine ions can "put out fires" within the human central nervous system as well; sedatives in the nineteenth and twentieth centuries included bromide for its calming effects. Doctors still sometimes prescribe potassium bromide for the treatment of epilepsy.

The industrial use of bromine, however, does have its drawbacks. Bromine, for example, plays a role in depleting the ozone layer, a phenomenon that scientists have studied extensively in the Dead Sea region, where the air naturally contains the substance, and, as a result, a natural ozone-free corridor has developed for a few miles just above the Dead Sea. Since this region lies much closer than the upper stratosphere, scientists take advantage of the proximity to learn more about holes in the ozone layer.

The murex uses bromine and indole to produce tyrindoxyl, which it stores in its hypobranchial ("under the lung") gland. This is the molecule that immediately precedes the dye. To get from precursor to dye, you need three components: an enzyme called purpurase, also present in the hypobranchial gland; oxygen; and sunlight. When you extract the gland from the snail and crush it in the open air, all the necessary pieces come together, and the precursor decomposes to form the final purple and blue dye molecules. As many as ten dye molecules can be formed from the precursors depending on the specific murex species, but the three most important are indigo (blue), monobromoindigo (violet), and dibromoindigo (red purple).[122]

A number of ancient sources indicate that the dyers of old believed that the snails had to be kept alive until the very last moment before they were used – clearly the reason that they were kept in pens and taken out on an as-needed basis. In a lengthy description of the dye process in his *History of Animals,* Aristotle comments: "Fishermen are anxious always to break the animal in pieces while it is yet alive."[123] Similarly, the Talmud records that, "the longer the snail stays alive, the more the dyer is pleased, so that the dye will be clearer."[124]

What the ancients learned by observation and experience, we can now explain biochemically. The enzyme purpurase, which facilitates the transformation of the precursor stored in the snail into the actual dye, decomposes quickly after the snail's death. The dye cannot be formed without the presence of purpurase, so completing the process of dye formation before the enzyme decomposes is crucial. In my own dyeing experiments, I have found that within a half hour after the snail dies, the brilliant color obtained begins to lose its luster and tends toward metallic gray. Only at this one stage – the extraction and crushing of the hypobranchial gland – is timing absolutely critical. The rest of the dyeing process can take place at a later time.

If the dye forms properly within the time limit, it proves remarkably stable and can last for years without any special care. Ancient texts also provide other information about dyeing processes. According to Aristotle, the murex yield the most dye in the springtime just before they gather to lay golden-colored eggs in large masses that resemble a honeycomb.[125] The correlation of abundance of dye with the egg-laying season points us toward a solution to a most perplexing mystery. Why do the murex store tyrindoxyl in their hypobranchial glands in the first place, and for that matter why do they produce it by adding bromine to indole to begin with?

A clue that may help solve the tyrindoxyl riddle has to do with the effect of the snail's sex on dye production. The difficulty in distinguishing males from females complicates experiments based on gender. Murex, perhaps due to pollutants in the sea, are often imposexual, a condition also called pseudohermaphroditism, in which females grow the murex's characteristic J-shaped penis – but still strangely remain female.

Not to be put off by such obstacles, Kirsten Benkendorff and her team at Flinders University in southern Australia have perfected a method of determining the current sex of the snails, an impossible task as long as they remain inside their shells. Somehow they have to come out of their shells and expose themselves. But how do you induce snails to shed their inhibitions and overcome their native modesty? Snails, it turns out, aren't that different from humans; plying them with drugs and alcohol seems to do the trick. Adding carefully calibrated amounts of magnesium chloride or ethanol to seawater in experimental tanks worked wonders. Once the snails were slightly drunk and totally relaxed, Benkendorff and her team gently pried them from their shells just far enough to get a peek at their little penises, thereby reliably distinguishing males from females.

Through painstaking research on the Australian murex species *Dicathais orbita,* Benkendorff has shown that females who lay the eggs produce more dibromoindigo than males.[126] The Australian team has shown further that the dye precursor molecule tyriverdin and a related compound, tyrindoleninone, exist in abundance in the egg masses of the murex and prove highly toxic to bacteria.

Putting all these clues together, we can propose a theory for why the murex produce tyrindoxyl and why they store it. The snails, primarily females, produce and store that molecule in the hypobranchial gland, which happens to lie next to the reproductive glands. In the course of the egg-laying process, the antibacterial tyriverdin transfers to the egg mass, providing the eggs with biochemical protection against infection.[127] What for the murex were reproductively advantageous biochemical processes and products became, for generations of humans totally unaware of and indifferent to snail biology, a convenient resource for the manufacture of beautiful purple and blue clothes.

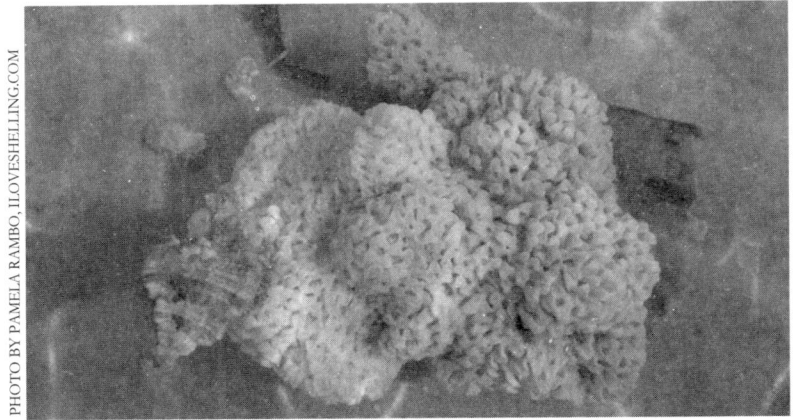

PHOTO BY PAMELA RAMBO, ILOVESHELLING.COM

Murex and egg mass

Using ethanol in the "sexperiment" also resulted in the secretion of mucus by the snails as they left their shells. This mucus contained Tyrian purple precursors, which scientists are studying for possible antibiotic and anticancer properties and uses.[128] Murex extract has long played a part in homeopathic medicine for treating disorders ranging from anxiety to cancer. Rigorous tests conducted by the Australians showed that, although the homeopathic murex remedy had little appreciable effect on cancer cells, material extracted from the murex egg capsules showed significant results. The substance destroyed many types of cancer cells or inhibited their growth; most promising was its effect on colorectal cancers, specifically the HT29 colon carcinoma, as well as certain lymphomas. Other molecules found in the snails, particularly indirubin, are undergoing rigorous study for use in treating different diseases, including Alzheimer's.

Scientists today are carrying out cutting-edge research investigating the properties of various mollusks for their medical or pharmaceutical value, but the methods developed and knowledge gained can also aid historians and archaeologists. We know that the composition and relative amounts of the different dye molecules vary across murex species, genders, locations, and seasons. As a result, those variables form a kind of fingerprint or signature. If you find a fabric dyed blue, determining the precise components of the dye can lead to an identification of the snail that produced it. For example, if examination of a bold blue pattern on a swath of ancient Chinese silk reveals one of our Mediterranean snails as the source of that dye, historians and anthropologists can draw significant conclusions. It is the task of a rare breed of professionals known as archeochemists to test ancient artifacts and discover their properties.

Archaeologists often work closely with historians to locate ancient sites that will yield interesting and important information, and then carefully dig to unearth relics and artifacts that will help them understand the story of the site, its historical timeline, and the people there. But this standard activity forms only part of the research that archaeologists can now pursue.

In order to understand fully what they have found and the implications of those findings, modern archaeologists carry out delicate chemical tests that can yield vital information, such as the precise age or composition of almost any substance or material. Such tests require great accuracy but use the smallest sample possible in order to inflict the least damage to the original object. Testing requires not only expensive and complicated procedures and devices, but also great talent and extensive experience. When it comes to determining the chemical components of ancient pigment, the tool of choice is an impressive instrument called an HPLC (High-Performance Liquid Chromatograph) , and no one uses it better than Israeli archeochemist Zvi Koren.

From his laboratory in the Shenkar College of Engineering and Design, Professor Koren, director of the Edelstein Center for the Analysis of Ancient Artifacts (named for Sidney Edelstein, the Chatanooga chemist), has examined dozens of samples from diverse archaeological sites. A scrap of a shroud from the Egyptian burial site of Beni Hasan; a tuft of wool from Masada; a blanket from the Cave of the Warrior in the cliffs overlooking Jericho, which may represent the earliest yarn-dyed textile ever found, dating back some six thousand years; a jug once belonging to King Darius of Persia – all have crossed his lab.

Although the arsenal of the archeochemist is formidable, Koren believes that high-performance liquid chromatography provides the most effective technique for determining the composition of ancient colorants, requiring only a minuscule sample – as little as a picogram, one millionth of a millionth of a gram –

while at the same time yielding the greatest amount of information.

First he dissolves the sample to be analyzed in a liquid – benzene, for example – and then introduces it into the machine at extremely high pressure. The liquid takes a path through a somewhat porous material and eventually exits on the other side. But as the liquid travels through the material, the different chemical components cling to the walls of the HPLC filter, each according to its own level of adhesion. That affects the time the chemicals spend in the HPLC, all of them taking different times to go through the machine. By measuring these various exit times, recorded as peaks on a graph with a time axis, chemists can determine not only what chemicals are present in the sample, but also their relative concentrations.

An analogy may help explain the process. Imagine a long gallery in a museum with many pictures hanging on the wall. A few paintings are modern, but most are classical. A large group of tourists enters the gallery. Counting them as they pass through the opposite door of the gallery, you find that they tend to leave in four groups. The first group has no interest in art at all and leaves quickly. A second group, which enjoys modern but not classical art, lingers at a few Kandinskys and Mondrians, then exits the hall a short while after the first group. Next to leave comes the largest group, those tourists with an affinity for classical art but not modern. Finally you'll have a small bunch of people who love both classical and modern art, and who looked at every painting in the room. That's how the HPLC works. Chemicals with the greatest affinity for the walls of the filter spend the longest time in the gallery, and you can figure out the size of the groups, whether chemical or sightseeing.

Using the HPLC, archeochemists can determine the ratios of the different molecules generated by various snails. Some produce primarily indigo or its brominated counterparts, others primarily indirubin or its mono- or dibromo- forms. When analyzing ancient purple and blue fibers, they look not only at the component molecules but also at the concentrations of each chemical, and in that way they can determine the type of snail that produced the dye.

French zoologist Henri de Lacaze-Duthiers had already noted that *Murex brandaris* and *Thais haemastoma* produce a red-purple dye, while *Murex trunculus* tends more to blue purple. Koren has developed a single index that accurately tells which snail species generated a given sample of dye. What he calls the di-mono index (DMI) calculates the ratio of dibromoindigo to monobromoindigo. *Brandaris* and *haemastoma* have high DMI numbers ranging from around 30 to the high 90s, whereas *trunculus* has a relatively low DMI, typically below 10. Analysis of the pigment found on that 2,500-year-old jug once belonging to King Darius of Persia showed a DMI of 5.6, so clearly that pigment must have come from *Murex trunculus* snails.

An HPLC filter can also analyze blue fabric dyed with indigo to determine whether the indigo derives from snails or plants. Murex indigo always contains some traces of brominated molecules, while plant-based indigo does not. Even if those trace elements are too faint to be observed by the human eye, the sensitivity of the HPLC machine will detect them. The Talmud records that only God could tell the difference between true *tekhelet* – indigo derived from the *hillazon* snail – and *kala ilan*, plant-derived indigo.[129] Modern science has now given us that capability as well.

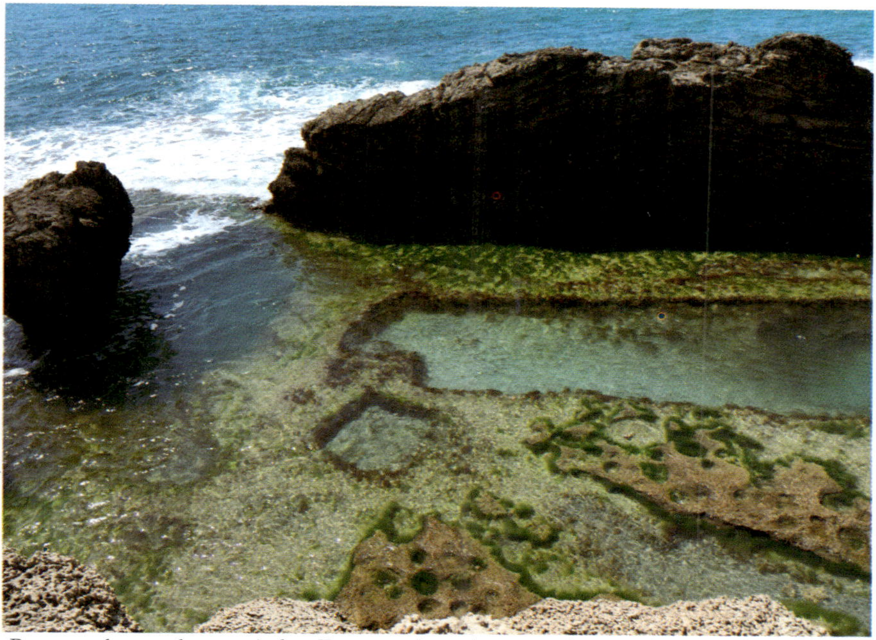

Baruch Sterman | tekhelet.com

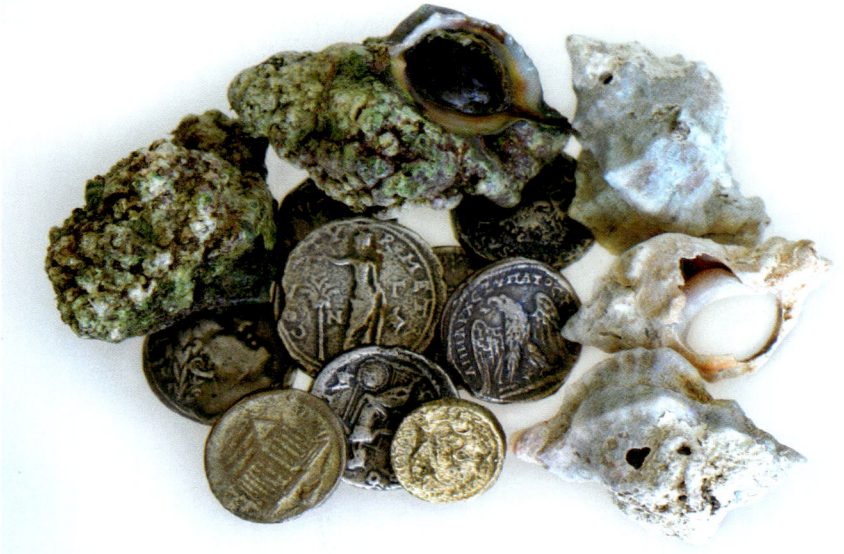

Rectangular pools quarried at Dor may have been used as holding pens for murex to keep them alive until ready for dyeing.

Photo by E. Weisberg

Ancient Roman coins depicting the murex snail, with actual murex shells.

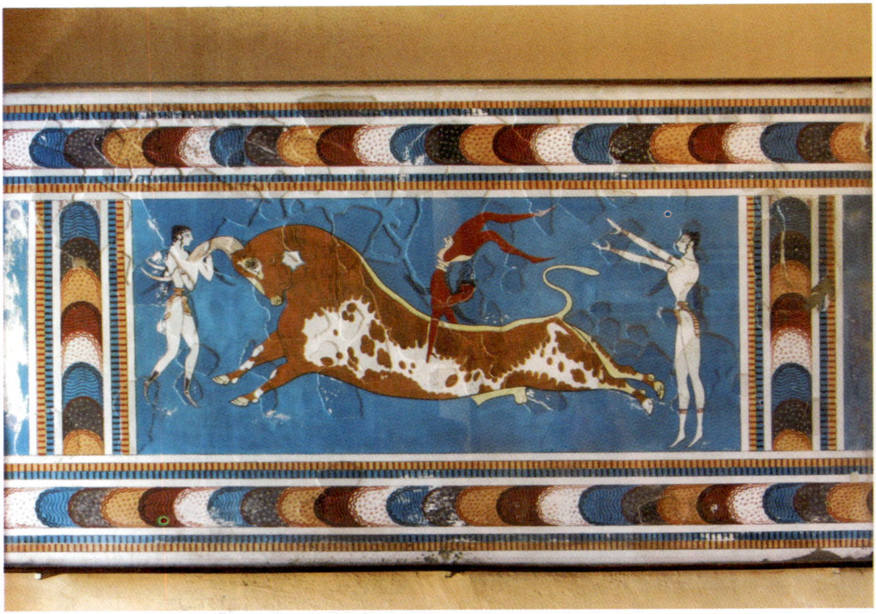

Hanna Viitanen / 123rf.com

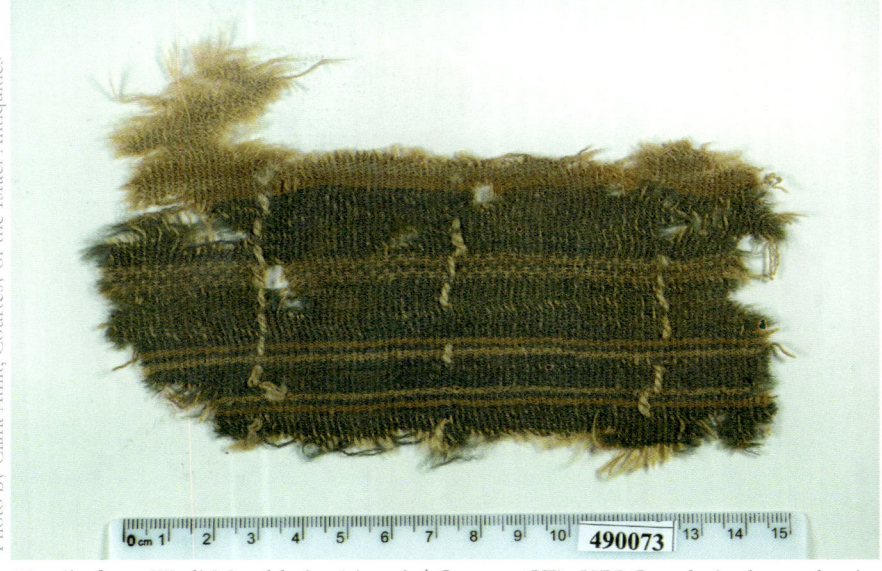

The bull-leaping frescoe from Knossos depicting the acrobatic maneuvers that may have been part of the Minoan cult of bull worship. The blue paint is apparently not of ancient origin, the restoration having been commissioned by Sir Arthur Evans, the archeologist who discovered the palace.

Photo by Clara Amit, Courtesy of the Israel Antiquities

Textile from Wadi Murabba'at (circa 2nd Century CE). HPLC analysis shows that it was dyed with blue from murex snails. Thread twists indicate it was manufactured locally.

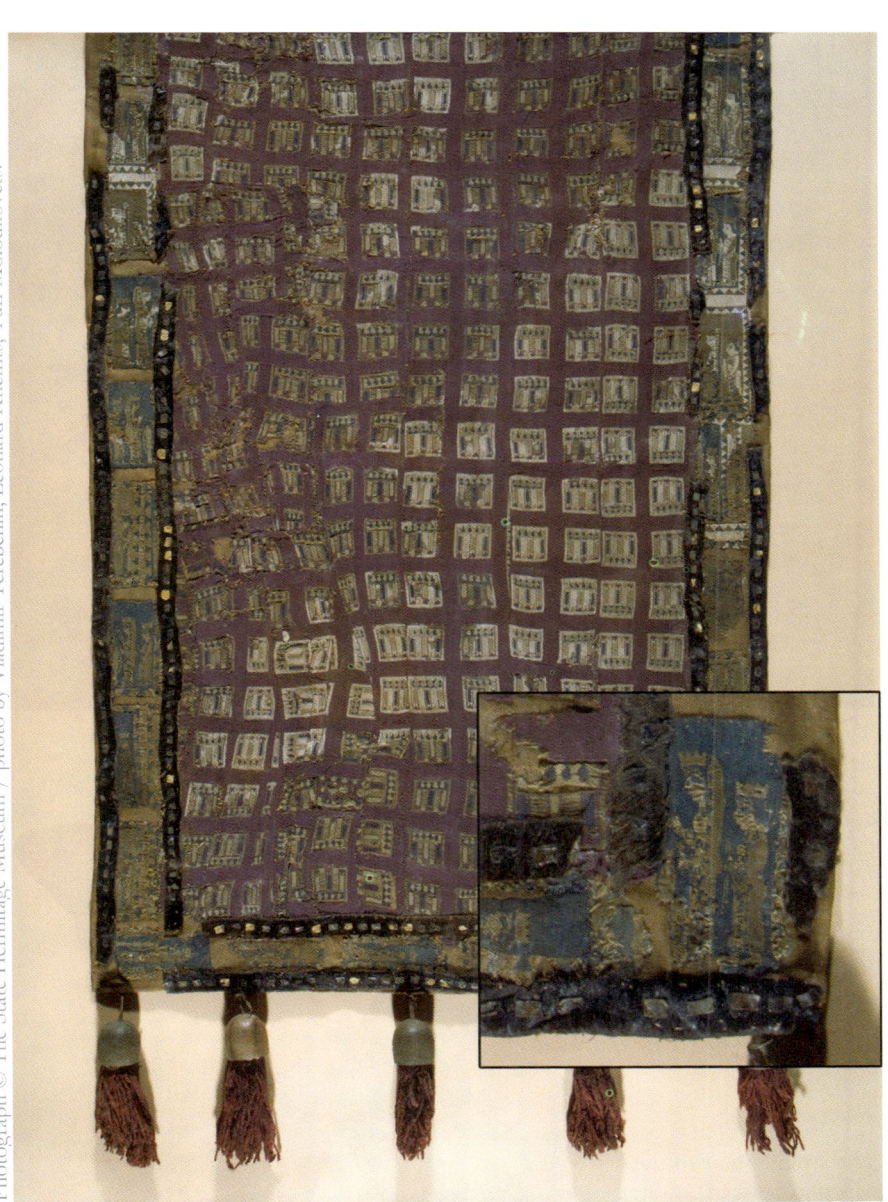

Saddlecloth from the Scythian burial kurgans of the Pazyryk culture (5th-4th century BCE), Altai mountains of southern Siberia. The presence of brominated indigo confirms that the purple and blue came from murex.

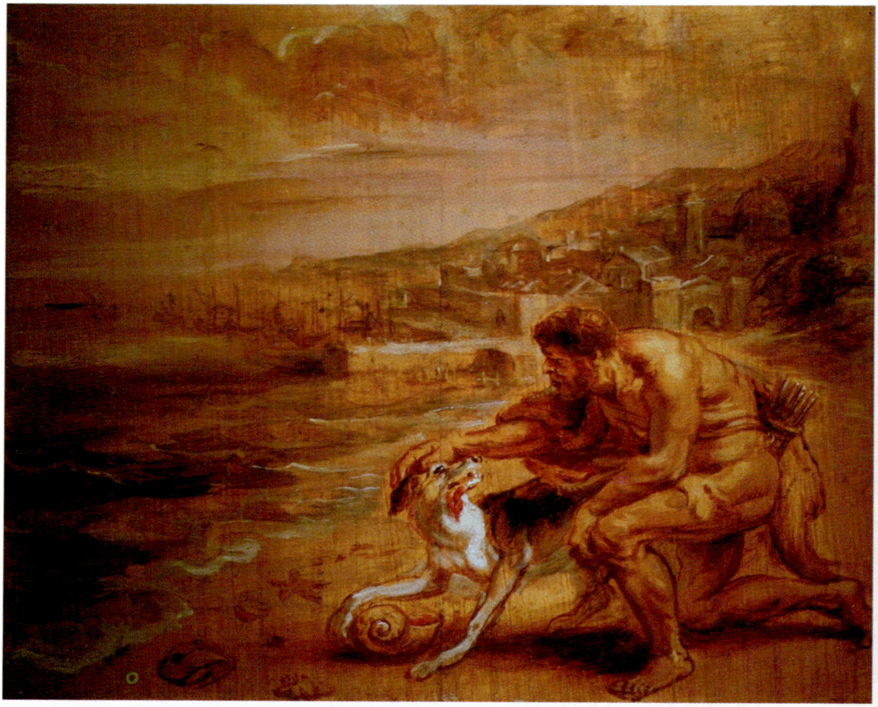

Wikimedia Commons

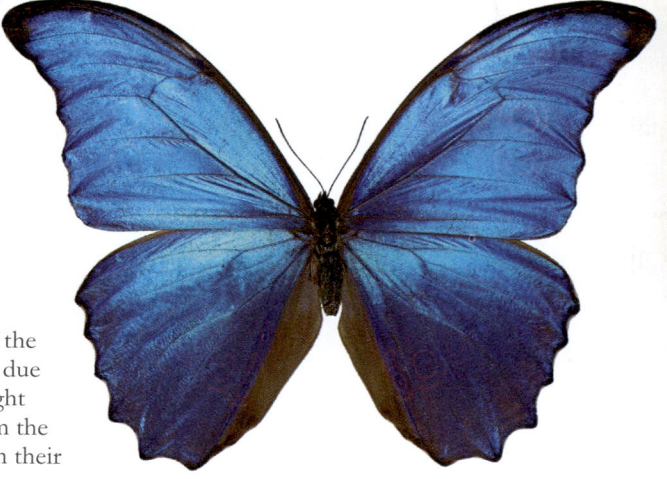

The legend of Hercules (Melkarth) and his dog discovering the purple dye, as depicted by Peter Paul Rubens, circa 1636. Musée Bonnat.

The striking blue of the Morpho butterfly is due to interference of light waves reflected from the ordered structures in their wings.

Lapis lazuli, a rare example of naturally occurring blue, was mined in Afghanistan
as early as the 4th millenium BCE.

Indigofera tinctoria, the source of indig

Isatis tinctoria, or woad.

Great Wave off Kanagawa, by Katsushika Hokusai, using Prussian blue to spectacular effect.

Sepia Officinalis, the common cuttlefish, source of the Radzyn *tekhelet* dye.

Various species of murex from the author's collection.

Ari Greenspan and the author on their first dive for murex snails, 1991.

Cracking open snails to obtain the dye glands.

Broken *Murex trunculus* shell exposing the hypobranchial gland where the dye precursors are stored.

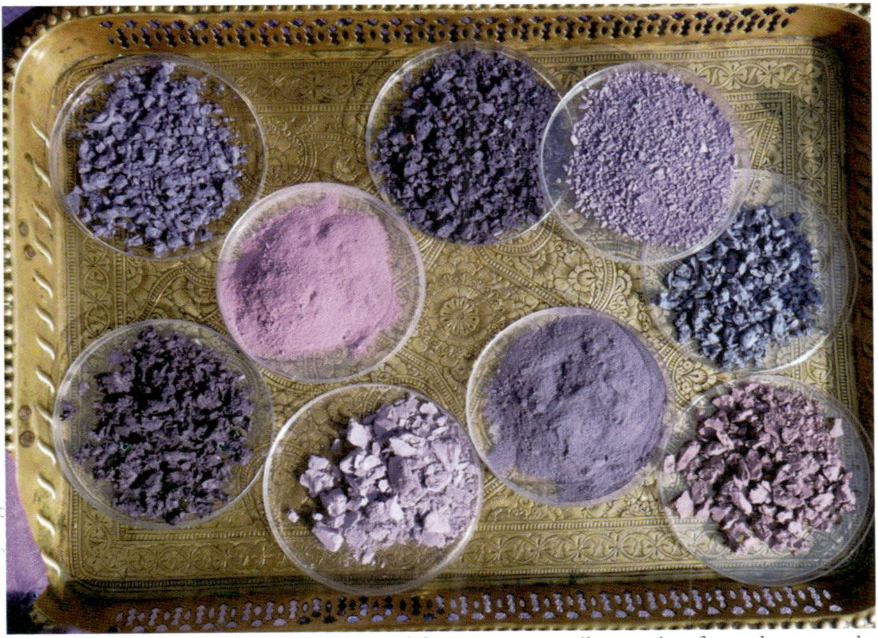

Photo by Inge Boesken Kanold

Assorted shades of dye powder derived from murex snails, ranging from deep purple to sky-blue.

Photo by E. Weisberg

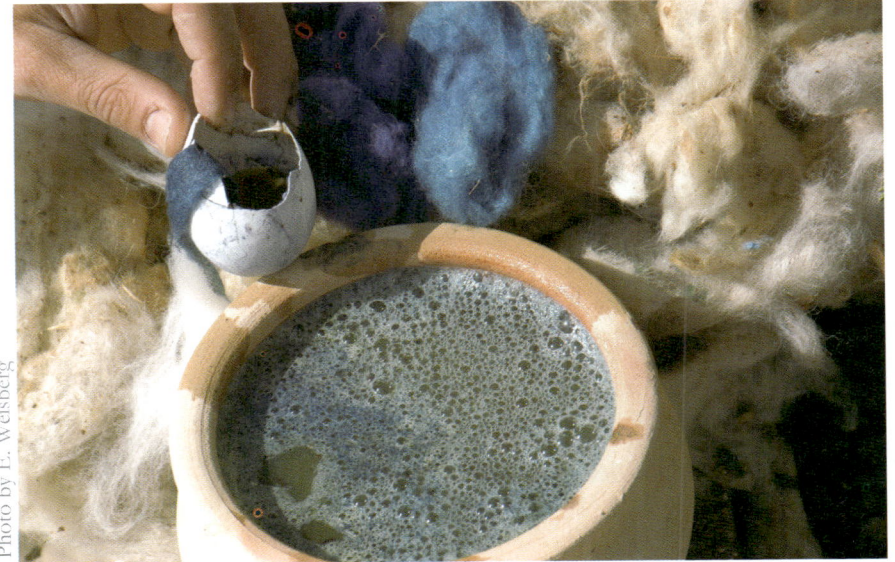

Sampling the dye solution in an egg-shell, as described in the Talmud.

Wool emerging from vat of Murex trunculus dye after exposure to sunlight. The dye at the bottom of the wool is still in the reduced leuco state while the blue at the top has undergone oxidation.

Asaf Stein | tekhelet.com

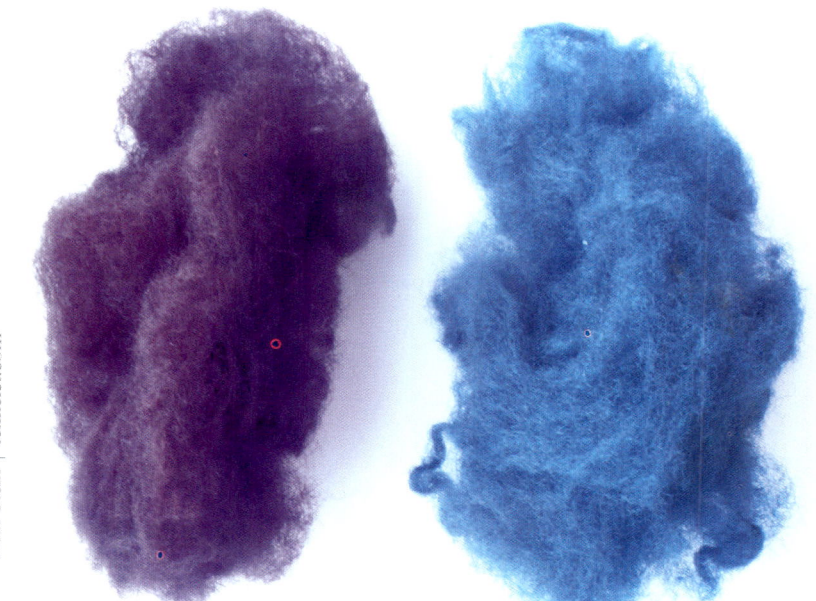

Tufts of wool dyed with *Murex trunculus* glands under identical conditions. The dye vat of the blue tuft was exposed to sunlight, the vat of the purple tuft was not.

Inge Boesken Kanold

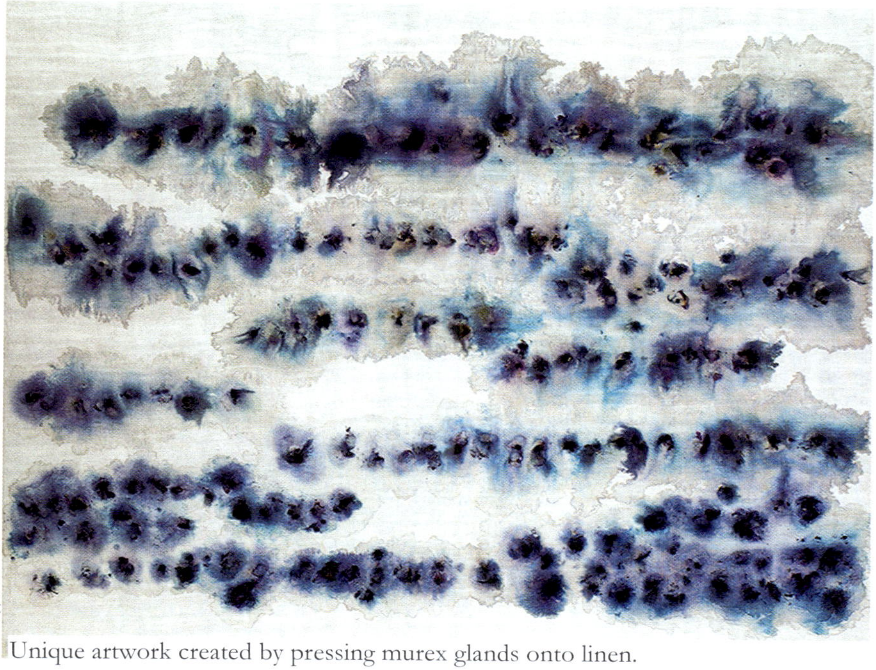

Unique artwork created by pressing murex glands onto linen.

Photo by E. Weisberg

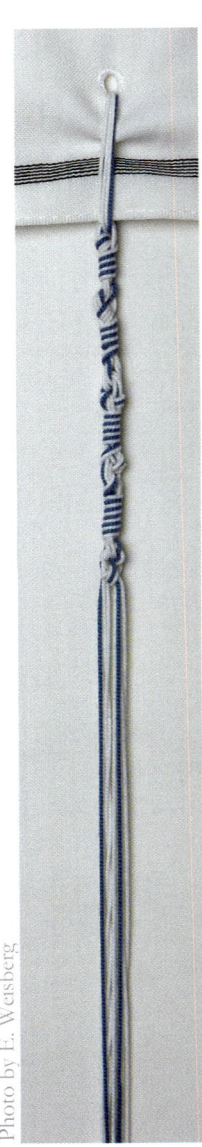

One method of tying tzitzit with *tekhelet*.

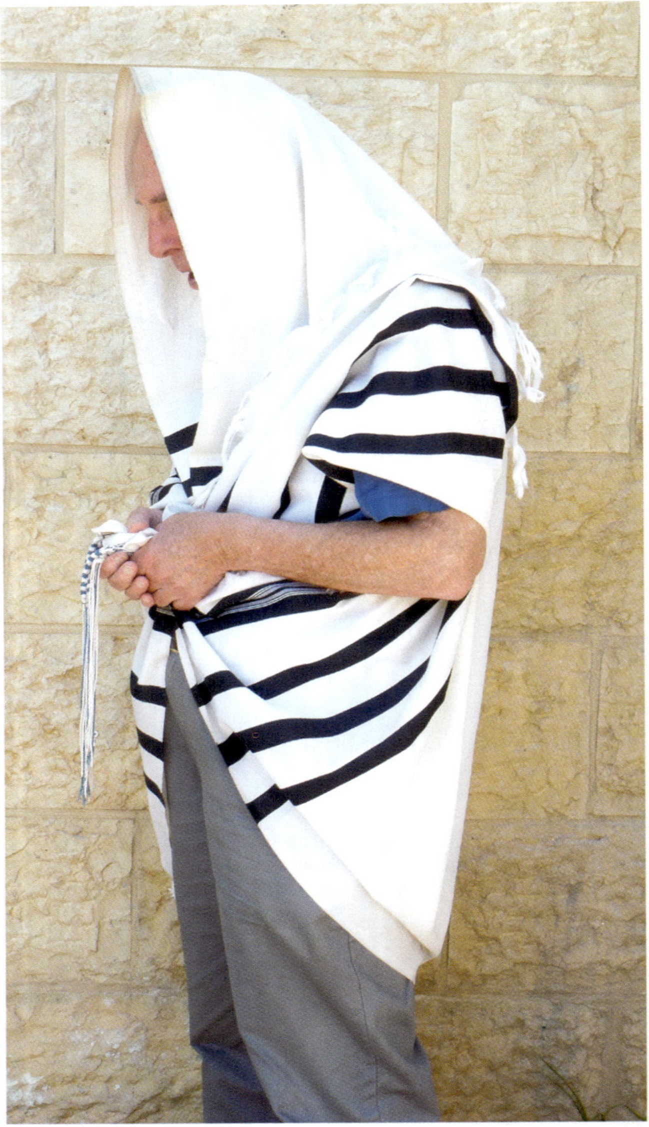

Jewish prayer shawl, tallit, with the *tekhelet* and white strings attached to the four corners.

The attire of the high priest in the Temple of Jerusalem. His tunic was made completely of *tekhelet*.

Mixtec dyeing in Oaxaca, Mexico

Above: *Plicopurpura pansa* secreting its dye.

Below: Traditional purple dyed Mixtec scarf next to *tekhelet* dyed with the Plicopurpura.

Left: Skein of freshly dyed cotton transforming to purple

Photos: Purpura Tradition and Sciende Foundation/Revelacion Huatulco Magazine/photos Nestor Diaz Diego

XI
DYED IN THE WOOL

$\mathscr{A}$thena, goddess of wisdom, taught mankind many useful skills – among them spinning, weaving, and dyeing – and she was patroness of those who practiced the crafts. The techniques involved in dyeing fabric, for instance, require such skill that it's no wonder that the ancients believed it beyond the abilities of mere humans to invent, coming instead by way of divine gift.

Actually, producing colored cloth isn't all that difficult; a glass of grape juice or a cup of espresso spilled on a white cloth will color it purple or brown, and after a few meals an infant's bib can easily rival a Jackson Pollock painting. But as everyone who has watched television knows, a good detergent will get out even the most stubborn stains.

In order to dye, however – as opposed to staining – a number of rare properties must converge. In addition to having a beautiful luster, the dye must be colorfast: It should not fade in sunlight or over time, and it cannot wash out with water or even soap. Dyed fabric must retain its colors permanently. In chemical terms, this means that dye molecules, when stable, must not bond readily with water or detergents. But therein lies the problem. If the dye molecule is inert, so that it doesn't bond to soap or water, it won't adhere to the fibers of a cloth for the same reason. The dyer must find a chemical ploy that will allow dye to enter the fabric, adhere to it, and get locked in so that it always remains there. An ancient dyer who could manage that trick wouldn't willingly share his recipe with others, and a certain mystique enveloped those who possessed the secret art of dyeing.

Dye masters of old commanded great esteem. The Jerusalem Talmud recounts that dyers walked around with a tuft of wool placed behind one ear, proudly advertising membership in their coveted guild.[130] Many cities throughout the ancient Mediterranean gloried in their skilled dyers, and trade in dyed fabrics played a major role in the economies of the times. Tyre in Phoenicia, where the much sought-after Tyrian purple was manufactured, is perhaps the best-known name in dyeing. Hermione, at the tip of the Argolis peninsula of the Peloponnesus, had one of the earliest dye centers in Greece, renowned for its purple dye stuffs that made their way as far as Persia. The island of Cythera – according to legend, the birthplace of Aphrodite – was named "the purple island" because of the cloth made from its native snails. Magdala, on the shores of the Lake of Galilee, the birthplace of Mary Magdalene, was sometimes called Magdala of the Fishes but was known also as Magdala of the Dyers. Thyatira in Turkey, site of one of the seven churches in the book of Revelation, was famous for its dye works, and Lydia, a cloth dealer converted to Christianity by Paul, lived there.[131]

Only the ancient Spartans, famous for their austere lifestyle, didn't allow dyers in their city because they felt that dyeing robbed wool of its true natural color. In the Spartan dialect of ancient Greek, the word for dyer was *dolun;* it was also the word for cheat.[132] Despite such principled opposition, the Spartans showed a more pragmatic side when it came to warfare: In battle they wore garments of red that effectively camouflaged any blood from their wounds.

While dyeing generated great wealth and repute for some, the actual manual labor involved in the process proved decidedly unpleasant. "The hands of the dyer reek like rotting fish, and his eyes are overcome by weariness," according to an ancient Egyptian papyrus. In Jewish law, a woman whose husband became a dyer after they married had the right to sue for divorce. She hadn't bargained on the vile stench that he brought home from the job.

Nevertheless, the vocation could confer status, and the highest prestige went to practitioners who worked with purple. In Roman times, they were called *purpurarii,* and the high standing of their profession appears prominently in epitaphs on ancient tombstones and sarcophagi. Only the *purpurarii* had the special skills, experience, and trustworthiness necessary to deal with the most precious of all dyes.

Stela of a Roman purpurarius
(purple dyer)
COURTESY OF ALINARI ARCHIVES, FLORENCE

Dyeing with murex isn't complicated theoretically, but in practice it presents considerable problems. Extracted and exposed to the air, the colorless liquid in the snail's hypobranchial gland quickly begins to turn colors, rapidly changing to dark purple. The secretion, if immediately applied to a cloth, will bond with the fabric and color it. Ironically, however, once the snail extract itself has turned purple, it can no longer bond with fabric. Dyeing can take place only while the substance remains in its colorless state – just a small window of opportunity. But in order to dye large amounts of material, you need a very large number of snail glands. One pound of wool, for instance, requires over seven thousand snails, and applying the dye one snail at a time is obviously impractical. The solution is to accumulate enough mollusk dye to color all the wool at one time. But collecting so many glands makes for a laborious, time-consuming process, and, when you finally have enough, all of it will have turned dark purple and therefore useless. Here's where the specialized skills of ancient dyers came into play. They discovered various recipes that could change unusable ooze back into a usable dye – a major scientific feat of enormous economic significance.

Dyes obtained from murex belong to a class known as vat dyes. In their natural state, such dyes don't dissolve in water, and it is difficult therefore to mix them into a liquid easily absorbed by a fabric. In order to bond with water, they first must undergo a chemical process known as reduction. Once reduced and in solution, the matrix of the cloth fiber can soak them up. At that point, when the dye molecules have thoroughly penetrated the cloth, they must undergo another process, known as oxidation, which returns them to their original inert state. No longer able to bond with water, they remain permanently fixed within the material – colorfast and fadeproof, impervious to sun and weather.

Reduction has many meanings in chemistry, but for our purposes we can define it as the gaining of an electron by a molecule, and its converse, oxidation, as the loss of an electron. When indigo or dibromoindigo molecules are reduced, their oxygen atoms become negatively charged, as can be seen in the accompanying diagram. In this reduced state, the molecules are called "white-indigo" or leuco-indigo (from the Greek *leukos* [λευκός], meaning white, bright, or clear), since in this form the dye liquid takes on a pale yellow or green color.

Dip a piece of wool into this mixture, remove it, and expose it to the oxygen in the air, and it loses electrons, undergoing oxidation. The yellow miraculously changes back to the deep blue or

X, X' = H, Br

Indigo reduction/oxidation reaction

purple of the initial substance. In medieval times, dyers who had mastered these amazing transformations were often thought of as magicians or sorcerers, revered and sometimes feared. *Der kann Hexen und Blaufärben,* "he can do magic and blue-dyeing" is a phrase still common in north Germany, which in olden days housed dye workshops.[133]

The gain of an electron that occurs during reduction adds a negative charge to the molecule, allowing it to bond to water. In this soluble condition, the dye can freely enter the densely packed fibers of the fabric. The vat dyers of old observed that of all fabrics wool yielded the best dye results. The negative charge, as it turns out, serves an additional purpose: It actually brings the dye molecule physically closer to the cellulose of the wool, facilitating a stronger chemical bond between the two.

Ancient craftsmen working with the leuco-liquid surely observed the effects of the bright Mediterranean sun on their handiwork, and the transformations brought about by sunlight enabled them to fine-tune the shades of the dye and therefore the fabric. From one basic liquid they could achieve a range of colors from deep burgundy to bright azure. As we've seen, the dye obtained from the murex consists of a mixture of indigo (blue) and mono- and dibromoindigo (purple and red purple). When the brominated molecules are reduced and in their leuco state, irradiation by ultraviolet rays, such as those present in sunlight, breaks the molecules' bond with bromine. As a result, the purple molecules become bromine-free and on oxidation yield a dye of sky-blue indigo.

The crucial process of debromination by sunlight, which creates our pure blue dye, occurs while the dye liquid still appears pale yellow; it cannot be detected until the wool is removed and fully oxidized. It was this invisible phase – known by the ancients and rediscovered in modern times by Professor Otto Elsner – that former researchers had missed. Rabbi Herzog, for example, convinced on the one hand that ancient sky-blue *tekhelet* derived from shellfish, couldn't accept the *Murex trunculus* as its source since he believed it could only produce purple.

Nowadays, dyers can accomplish the reduction reaction quite easily by means of a strong reducing agent, such as sodium dithionite. In ancient times, however, the process required extreme care and sensitivity, and the discovery of the method for reducing the pigments obtained from the murex into true dyes represented one of the greatest – and therefore most guarded – technological achievements of the ancient world. In his *Natural History,* Pliny the Elder describes the complexities of the dye process:

> *Subsequently the vein of which we spoke is removed, and to this salt has to be added, about a pint for every hundred pounds; three days is the proper time for it to be steeped, as the fresher the extract the stronger it is, and it should be heated in a leaden pot, and with fifty pounds of dye to every six gallons of water kept at a uniform and moderate temperature. This will cause it gradually to deposit the portions of flesh which are bound to have adhered to the veins and after about nine days the cauldron is strained and a fleece that has been washed is dipped for a trial, and the liquid is heated up until fair confidence is achieved. A ruddy color is inferior to a black one. The fleece is allowed to soak for five hours and after it has been carded is dipped again, until it soaks up all the juice.*[134]

A briefer, less-detailed description appears in the Talmud, which records a conversation between Abaye, a fourth-century Babylonian scholar, and Rabbi Samuel of the Land of Israel:

> *Abaye enquired of Rabbi Samuel, the son of Rabbi Judah, This tekhelet, how do you dye it? He replied: We take the blood of the hillazon together with other ingredients and put them all in a pot and boil them together. Then we take a little of the liquid out in an egg-*

shell and test it on a piece of wool; and we throw away what remains in the egg-shell and burn the wool.[135]

Both of these descriptions share the notion that the dye needs to be tested in order to determine if it is ready for use. Generations of Talmud students have wondered about the insistence that the dye be sampled; if you want to know the color of the dye, why not just look at the liquid in the vat? The answer, of course, as we now know, is that the mixture must first be reduced to its leuco state, in which the color appears completely different from that of the final dye. The only way to know the actual shade that the fabric will eventually assume is to use some wool as a dipping material and expose it to oxygen – hence the Talmud's instruction to take a bit of the dye in an eggshell, the ancient version of a disposable cup.

Pliny's and other ancient recipes call for the dye vat to stand for many days over a low flame. This instruction has led modern researchers to the assumption that ancient dyers achieved reduction, transforming the murex dyes into their leuco state, via fermentation. The prolonged heating and steeping of the dye provided the perfect conditions for fermentation, a process performed by microorganisms such as yeast or bacteria that would thrive in the warm vat of snail soup. Bacteria stewing in the vat for a week or two presumably accomplished the same function that modern chemical reducing agents can achieve in seconds.

The precise details of the ancient fermentation process, however, eluded modern biochemists. As recently as 1987, researchers Otto Elsner and Ehud Spanier remained uncertain as to how the ancient dyers "were able to reduce the dye and keep it in reduced state during the preparation and dyeing procedures that lasted… many days."[136] A related mystery also remained unanswered: Which specific bacteria were actually responsible for the fermentation of the snail dye? Ironically, it wasn't pedigreed scientists in the respectable laboratory facilities of some university who solved these riddles, but rather an amateur chemist in the back of a shed at a children's museum outside London.

After twenty-five years as an industrial engineer, John Edmonds retired at the age of sixty-one. His passion had always been medieval England, and over the years he had become a world-class expert on Chaucer, having produced the only modern English translation of the author's complete works.[137] An active member of the Medieval Dress and Textile Society, he spent much of his retirement volunteering at the Chiltern Open Air Museum, a lovely place where visitors can get a taste of medieval and Victorian England.

Edmonds developed a strong interest in one of the museum's main attractions, the natural woad-dyeing exhibit, and through his knowledge of medieval history and life he helped make the presentation as authentic as possible. In 1995 the *Daily Telegraph* ran an article about John's research into woad dyeing by means of natural fermentation. Someone in London read the story and sent it to my friend Joel Guberman in Israel. Joel contacted Edmonds and asked if the natural fermentation process that he had applied to dyeing with plant-based indigo might shed light on the ancient methods of dyeing with shellfish-derived dyes.

This question piqued John's curiosity, and he asked Joel to send him some murex dye so that he could experiment with it.

Ever the scholar, Edmonds read the literature, both ancient and modern, relating to murex dyeing and fretted over the problem of how the ancients reduced the murex pigment in order to obtain the leuco form required for dyeing. He knew of Lacaze-Duthiers's famous encounter with the fisherman who stained his shirt purple by smearing it with snail guts. On the basis of this experience, Lacaze-Duthiers had suggested that in ancient times dyers did their actual dyeing with the precursor, which exists inside the snail in a reduced form, taken directly from the live snail's gland and immediately applied to the wool.

Edmonds rejected that idea. The ancients established a vast industry that would have required mass production and also, to use modern terms, quality assurance and protection of intellectual property. Dyeing wool snail by snail, Edmonds reasoned, would have required a truly massive workforce. Even if such a large team could have been assembled in antiquity, the quality and consistency of the resulting colored wool, each murex giving a slightly different shade or hue, would never have met the demands of the fashionistas of Athens and Rome. Furthermore, as Edmonds points out, those controlling the dye industry would have wanted to keep the number of laborers to a minimum in order to keep trade secrets closely guarded and to reduce the chances of industrial espionage. Because such practical considerations argue against dyeing snail by snail, Edmonds wondered if any of the historians and scientists who had discussed the murex had ever had any real-world experience in dyeing.

One detail in Pliny's recipe caught Edmonds's attention: the "fifty pounds of dye to every six gallons of water." The term *dye* here couldn't mean just the pure distilled extract; if it did, that mass of liquid derived from tiny glands would have required approximately three hundred *million* snails. There must have been a great deal of snail meat mixed in with Pliny's "dye." Edmonds believed that the dyers carefully extracted the glands of larger snails, but they probably just crushed the smaller ones and threw them whole into the vat.

Searching Pliny's text for more clues, Edmonds brooded over the salt – "about a pint for every hundred pounds." He surmised that the level of salinity in Pliny's mixture wouldn't have adversely affected the bacteria he was seeking, which used the soft snail flesh as their growth medium to carry out the fermentation. To prove his hunch, he conducted an experiment using the murex extract that Joel had sent. But that dye was dried and ground into a powder, and Edmonds needed fresh snail meat. At a local market, he bought a jar of pickled cockles. Saltwater clams, he thought, should be essentially the same as murex meat as far as the bacteria were concerned. He took an old two-pound jam jar and set up his miniature dye vat. He washed the vinegar off six cockles and placed them in the jar along with the dried murex. He added water and salt, chemically adjusted the pH of the liquid to a moderately alkaline nine, placed the jar in a water bath, and fixed the temperature to a cozy fifty degrees Celsius – the same conditions that he had used for dyeing woad – keeping the environment constant. Over the course of ten days, the purple color of the mixture slowly turned to pale green. Edmonds dipped a piece of wool in the brackish liquid and left it immersed for a few hours. As he removed the wool and exposed it to air, the color turned from green to a dark reddish purple. John Edmonds had just conducted the first instance of purely natural shellfish vat dyeing in over a thousand years.

But what about his search for the specific bacteria involved in the fermentation process? I met John a few years later, and we talked in depth about his research as we strolled along the leaf-strewn paths of his beloved museum. The bacteria that played the starring role in this drama, he told me, belong to a genus called *Clostridium,* stronger than most bacteria. The salt that Pliny mentioned wouldn't have killed them, but it would have suppressed the growth of different bacteria that might otherwise have interrupted the dye process. Clostridia, it turns out, are nasty little bugs that come in dozens of varieties, each with its own impressive Latin name. They are responsible, among other things, for certain forms of colitis and gangrene, and for the neurotoxins that produce tetanus and botulism. *Clostridium botulinum,* however, has had a more glamorous career in recent years in the form of Botox, the popular antiaging alternative to plastic surgery.

Like other microorganisms, clostridia help with bacterial fermentation, such as the fermentation of acetone, butanol, and ethanol from starch. *Clostridium acetobutylicum,* known as the Weizmann organism, takes its name from Dr. Chaim Weizmann, who used it to develop the process and is recognized as the founder of industrial-scale fermentation. During World War I in England, he helped produce large amounts of acetone used to make cordite, the indispensable substitute for earlier gunpowder. Weizmann's ardent Zionism and his vital role in the British war effort, as well as his long association with Foreign Secretary Arthur Balfour, played a key role in the 1917 Balfour Declaration supporting the establishment of a Jewish homeland in Palestine. Weizmann became the first president of Israel in 1949.

John Edmonds's experiments with bacterial fermentation were a matter of pure intellectual curiosity, but they came to the attention of more pragmatic and lethal-minded people. "The work that I do here is partially funded by the UK Ministry of Defense," he told me. When they heard he was doing research that involved *Clostridium,* they wanted to know everything he learned to see if his findings had any implications for biological warfare. John died a while later at a relatively young age. The scientific community lost a renaissance man in every sense of the term.

Edmonds had followed Pliny's instructions in a search for historic authenticity. The historical and scientific record concerning murex dyeing had always fascinated me, of course, but my main interest from the beginning was essentially practical. I wanted to create woolen strings colored blue with the dye obtained from *Murex trunculus* snails, and together with a few friends who shared my enthusiasm, I eventually formed an organization to produce *tekhelet.*

Catching the snails in the Mediterranean is not as easy as it sounds. Underwater, it's hard to tell the difference between the snails themselves and the gnarled, grayish rocks on which they live. Moss and sea fouling cover them both, making them virtually indistinguishable to the untrained eye. Over the years, we have become experts at spotting the little creatures, and after an hour or so underwater we can collect a good few hundred snails. Back on shore, we crack them open with a hammer, extract the small yellowish glands using a utility knife, and throw them into a jar. On exposure to air and sunlight, the slimy contents quickly turn dark purple.

When we first started, the rest of the procedure usually took place in my home. I came back from a day at the beach with a jar of snail glands, ran up to our balcony overlooking the hills of Jerusalem, and got started. I had long grown used to the pungent, garlicky smell mixed with the reek of rotting fish that emerged from those jars – but my family never did. During her pregnancies, my wife often became nauseated when I walked into the house with my purple treasure – a reaction that other men might have found a bit disconcerting.

With a handheld electric mixer, I blended the glands until they achieved an even consistency, then spread the purple sludge onto a tray. The tray went on an old hot plate from my college days, and I covered it with a strip of window screen to shield it from flies. The paste dried out on the hot plate for a few days until the flat, thin sheet snail gland goop had hardened fully. Next, I broke the sheet into small pieces and processed them in a coffee grinder until they became a fine purple powder. Extremely stable, this powder, or "instant dye," can be stored without refrigeration for years. Some five hundred snails weighing a little over twenty pounds can yield approximately thirty grams of purple powder, enough to dye about a third of a pound of wool.

As I developed more expertise and as our operation grew, we eventually moved to a small workshop – much to my wife's delight – where we continued the rest of the dyeing procedure. That location remains our headquarters today. We prepare the dye solution by placing a teaspoon of the powder into a beaker of hot water and then stirring caustic soda into the mixture. Caustic soda, also known as sodium hydroxide or lye, is a very strong base typically used for cleaning out stopped-up drains. It does this by dissolving the proteins, fats, or other organic material that usually clog pipes, and it serves essentially the same function in the beaker, as well as providing the alkaline environment necessary for the reduction to take place. The purple powder obtained from grinding the dry, hardened sludge isn't clean, pure dye powder; it also contains the ground-up bits of meat, fat, and other impurities from the snail. The caustic soda dissolves those unwanted components but has no effect on the inert, resilient dye molecules.

After a few minutes, we add a bit more hot water, along with sodium dithionite, which reduces the dye, changing the liquid to the yellow color characteristic of the leuco state where the dye fully dissolves in water. We pour the mixture through a stainless steel strainer into another beaker, and the sediment, bits of shell, and other debris filter out, leaving a clear fluid. This fluid – the end result of all our labors – then goes outside into the bright sun. There, over the next hour or so, the sun's ultraviolet rays do their debrominating. (Remember, we don't want purple, so we want to end up with as much indigo in the mixture as possible.) Now the dye can do its appointed work.

We place only the finest wool, shorn from merino sheep, onto a carder, a wooden brush with fine metal teeth. We gently scrape another carder across the first, and, as the wool catches between the teeth, the fibers break, separate, and straighten. Peeling the clean wool off the carders, we gather a few grams into a tuft to be immersed in the dye vat. Before doing that, however, we must adjust the dye mixture.

If you introduce fine wool into the caustic soda's very high alkalinity, the wool would be damaged and turn to felt. Adding ammonium sulfite, a mild acid, brings the pH down to about eight. Now we can safely dip the wool into the beaker and agitate it gently from time to time to ensure that the liquid penetrates fully for an even dye.

After soaking for about a half hour, the wool goes from the vat into the air. As the leuco dye slowly oxidizes, the miraculous transformation begins. Watching the color spread throughout the wool, gradually turning the yellow to sky blue, is as mesmerizing today as it was the first time I saw it over twenty years ago. In our workshop, we have successfully produced a modern version of an ancient dyeing process to create the blue wool prescribed for ritual use in the Bible.

To make the tzitzit for the Jewish prayer shawls, the wool must then be spun into threads. This we do with an old-fashioned spinning wheel, which produces thin threads twisted together to end up with an eight-ply blue cord. We wind the eight-ply string a number of times around a simple, handheld H-shaped wooden apparatus delightfully called a niddy-noddy, used to make skeins of specific lengths of yarn or thread. With the strings wrapped tightly around it, the niddy-noddy goes into a steam bath, which sets the wool and stops the string from unraveling. Here we discovered another mystery as yet unexplained by chemists.

According to the standard understanding of these types of vat dyes, the only way to turn the murex dye to pure sky blue is to expose it to sunlight while in solution, in its leuco form. The ultraviolet rays debrominate the molecules and leave pure indigo in the solution. Once the treated wool undergoes oxidation, it has received its final color, whether blue or purple, and will not change. Indeed, we have dyed without sunlight and, as expected, produced purple-blue strings. However, we've found another way to change that purple to blue after the strings have already undergone oxidation.

When purple strings go into the hot steam bath, they lose their violet hue and change to sky blue. This happens most of the time, but the change isn't 100 percent consistent. In order to ensure pure blue strings, without any hint of violet, we don't rely on steam but, rather, expose the leuco liquid to the sun. The behavior of dyed wool in hot steam and the parameters of that action – for example, the length of time after dyeing in which this effect takes place – will make an intriguing subject for some doctoral student in chemistry one day.

Using modern techniques and chemicals makes dyeing today far more efficient than it could ever have been in ancient times. We are constantly looking to improve our production methods to achieve greater yields from the snails, since sustainability and other environmental concerns are also important factors. That is why I was intrigued when I received an email from Kathy Leyva-Gomez, a school teacher in Chicago, who had heard of our work. Kathy traced her family roots back 500 years to *Conversos* who had fled Spain and the Inquisition to the promise of safety in the New World. They settled in the hills of Oaxaca, Mexico, in the small town of Pinotepa de Don Luis, which happened to be home to a large population of Mixtec Indians, an indigenous people who had been living in that region for thousands of years. One of the most noteworthy features of Mixtec culture is their use of sea-snails to dye beautiful purple fabrics for commercial as well as ritual purposes. Kathy's family still lived in Oaxaca, and she asked if I would be interested in joining her on her next visit to meet the Mixtec dyers and to learn about their methods.

Beyond my sense of adventure, and curiosity about an exotic culture and a fascinating history, and even apart from the practical and ongoing search for additional supplies of *tekhelet* dye, the trip held a special fascination for me. I was aware that over the centuries the Mixtec dyers had perfected a unique method of "milking" their snails – *Plicopurpura pansa*, a species found only along the Pacific coast of Mexico and Central America – and could obtain the dye from the same snail multiple times, sparing them so that they could live – and dye – another day. Might this be a technique that we could potentially employ in our *tekhelet* dyeing? I replied to Kathy that, of course I would love to meet the Mixtec.

Traveling from Israel to Oaxaca, (a long and exhausting journey that took me more than two full days and three cancelled flights) I was reminded of that arduous trek taken in the service of *tekhelet* almost 130 years ago by Rabbi Gershon Henokh of Radzyn. Though my trip was longer in distance, his excursion by horse-drawn cart would have been much more dangerous, with discomforts far greater than a protracted airport layover.

I landed in Huatulco, a sleepy little town with gorgeous natural coves and lagoons, and magnificent cliffs overlooking the endless Pacific. At the airport I was greeted by Kathy and by Marta Turok, an anthropologist who had worked with the Mexican government. Marta had spent decades researching the Mixtec peoples and became a pivotal force in the enactment of laws and regulations focused on protecting the snails, their habitats, and the culture of Mixtec dyeing. They had prepared a packed agenda for the week, but the most important part for all concerned was working with the dyers and the snails, and seeing if the purple dye produced by the *Plicopurpura pansa* could indeed be used to make blue *tekhelet*.

We got to work almost immediately, going directly to one of the beaches in Huatulco where I met the rest of the team. Delia Domínguez-Ojeda is a marine biologist at the University of Nayarit, Mexico, who has spent many years researching the Plicopurpura, their life cycle, anatomy, and environment. Don Habacuc Avendaño Luis is the oldest and most experienced of the nine remaining traditional Mixtec dyers still practicing the ancient craft. At 75 year old, Habacuc has been dyeing his entire life and knows the snails intimately, their characteristics and habits, their environment, their likes and dislikes. Habacuc's ancestors have been coming to these rocky shores of Huatulco, in the Oaxaca region of southwest Mexico for countless centuries.

The first day we made a preliminary trip to a local beach to meet the snails themselves and to experience firsthand the dye secretion. The Plicopurpura adhere to rocks just above the waves. I watched as Delia gently pulled one off and then spit on it. A yellow liquid started to fill the shell's aperture. "Sometimes you have to irritate the snail," Delia explained, "and nothing irritates a snail more than someone spitting on it." I let it ooze out onto my hands and watched as over a few minutes the yellow turned purple. This was not surprising to me. The dye from our Mediterranean Murex snails undergoes the same transformation as we break them open and extract the gland, which is covered with a similar yellow paste.

As I expected, the purple stained my fingers, and I knew it would take a few weeks to fade. I smelled my hands and recognized the familiar pungent garlic-like odor. Habacuc watched what I did, and I mentioned (with Delia translating) that our snails back in Israel had the same smell. He was excited to hear that and told me that when he comes home from dyeing, his family complains about the smell. I exclaimed that my family feels the same way, and we both laughed. It was an interesting moment: two men from opposite sides of the world, from cultures and backgrounds poles apart, with no common language, yet connected in some deep way through their shared appreciation for the wonders of the magnificent dye-producing snails, he for his reasons and I for mine.

My first encounter with the Plicopurpura encouraged me, since the characteristics of its secretion indicated that these were comparable to the yellow substance we obtained from our snails. The real test would be to see if the excretion behaved the same way during exposure to sunlight and if it indeed could be transformed into *tekhelet*.

That night Marta and Habacuc told us more about the Mixtec and their purple dyeing. Traditionally, only men may do the dyeing, and it is forbidden for women to touch the snail, called *T'shiinda Ya-A* in Mixtec. The dyers – *Ra Yaki Yuba-A*, literally the men who dye threads – would come down from Pinotepa de Don Luis in the mountains to the coast of Huatulco, a trek that took up to 12 days before there were cars, laden with as much cotton yarn as they could carry. Then they would spend their days dyeing the threads, and working in the local fields in return for food and shelter. After three months or so, when all the cotton was finally dyed, they would return to the village, where the women would weave the purple threads into scarves, shirts, and especially skirts. The most beautiful and meaningful of the skirts is called a *Che-Eh*, worn by a bride at her wedding, after which it is set aside, never to be used again until her death, when she is buried in it.

I was struck by the symbolism of the purple color, which the Mixtec associated with blood – the blood of childbirth, fertility, and death. In the ancient Mediterranean and Near Eastern cultures, the association of purple and blood came to represent victory in battle. Though purple was prized above all other colors, both by the Mixtec and the ancient Mediterranean and Near Eastern cultures, the Jews esteemed the sky-blue *tekhelet* as the most valuable and holiest of all colored fabrics. Purple, though beautiful, is earthy, but *tekhelet* soars above and reaches to infinity.

The following day we set out early by boat under Habacuc's direction to a secluded cove that held a very large number of Plicopurpura. Marta explained that the Mixtec, who have been dyeing for centuries or maybe millennia, adhere to four principles that have helped keep the snail population healthy and strong. First, they do not touch snails that are mating, so as not to interrupt the reproductive process. Second, they do not use juvenile snails. Third, they do not come back to the same cove until at least one month has passed. And last, they take care not to harm the snails and to gently return them to the rocks after 'milking'.

I followed Habacuc over the slippery rocks in search of the snails. He would take each snail and lightly press it against the cotton threads, which absorbed the yellow liquid. As the day progressed, the skein of yarn changed colors from shades of yellow to green and eventually to its final purple. I was busy meanwhile collecting the liquid from each snail in a bottle, one by one, drop by drop. Habacuc and I worked together for close to six hours under the hot equatorial sun. After milking close to a hundred snails, I had about a quarter-cup of liquid, which had turned a

Habacuc the Mixtec, and the author searching for Plicopurpura in Huatulco
COURTESY OF PURPURA TRADITION AND SCIENDE FOUNDATION/REVELACION HUATULCO
MAGAZINE/PHOTOS NESTOR DIAZ DIEGO.

rich dark purple. Habacuc was annoyed with me, since as far as he knew, once the snail discharge turned purple it would no longer adhere to the threads, and he felt that I had wasted all that dye. But I knew that the ancient Mediterranean dyers who broke open their snails and collected the glands had learned to take that purple and process it so that it could adhere to a fabric and be used for dyeing.

Back in Huatulco, I set to work on the final and crucial part of my experiment, and I was nervous and unsure if it would succeed. Were these Mexican snails really similar to their Mediterranean counterparts, and would the dye obtained from them have identical properties and characteristics? My conditions were far from laboratory grade, as the dyeing would take place outside over a chafing dish burner that we got from a nearby restaurant, and I would be mixing in chemicals with no accurate way of measuring the amounts or knowing the concentration of the dye within my liquid. It was also late in the day, and I was worried that the sunlight might not be strong enough to effect the changes from purple to blue.

The entire process of reducing the purple liquid, exposing it to sunlight, and immersing the wool, took about 45 minutes. With our entire team looking on, I finally extracted the tuft of wool and waited. Slowly, magically, the color began to transform. First bright yellow, then green, then grey, and finally a majestic, gorgeous blue – so similar to the Pacific Ocean behind me and the tropical sky above. For the first time in history, *tekhelet* had been obtained from a snail outside the Mediterranean, and without a single creature being harmed.

Fascinating as the Mixtec process is, it is not presently something that we can look to implement in any practical manner. Furthermore, in terms of Jewish Law, it is not at all obvious that the *Plicopurpura pansa* would be acceptable as a valid source for *tekhelet* dye. After all, the ancient Israelite dyers used the *Murex trunculus*, and though the two species are related (both are members of the family Muricidae) and the dye they produce are apparently similar in constitution, they are not identical. If one requires authenticity, then only the *trunculus* would be legitimate. This question is one best left to Rabbis and scholars to debate.

But there is, nonetheless, a very important lesson to learn from the encounter with the Mexican dyers. They have a deep respect for the welfare of their snails, and do all they can to protect that valuable resource upon which they so profoundly depend. We must do no less with our precious resources and have implemented a number of policies to that end. First, we spread out the area from which we collect our snails over several different locations each year, and stagger those locations from year to year. Smaller juvenile snails caught in the nets are thrown back. We also take samples of the snails each year and monitor the size and weight to ensure that the Murex populations remain healthy. Finally, we must work with local environmental groups to keep the oceans and coasts clean and pollution-free.

XII
BEHIND BLUE EYES

$\mathcal{M}$arvelous photographs of Earth taken from space reinforce the aptness of the description of our world as a blue planet. If we shift the perspective, however, and revert to our earthbound viewpoint, the picture changes radically. Blue is actually extremely rare. Most of the natural landscape we encounter consists of earth tones, shades of red, beige, and brown, interspersed with the various greens of vegetation. Aside from the sky and bodies of water – which are never uniformly blue all the time – our biosphere largely lacks the color. There are, to be sure, some blue flowers and birds, the occasional frog, butterfly, or gemstone – but even the blueberry is more purple than true blue.

It seems that nature conspired to produce a world with a minimal amount of blue. Is it a random oddity that has no particular cause? The fact is that the laws of physics impose limits on the generation of blue, and those laws are unyielding. They permit only five natural ways to make blue, and of those only one can act as the basis for the biological processes that lead to the formation of a blue dye molecule.[138] The first four produce the blue of the sky, of some birds' feathers, of the ocean, and of certain precious gems. The last process is responsible for almost all the color in our world, from green grass and yellow bananas to red flowers and brown rocks. But it cannot produce blue. The immutable laws restrict the possible range of colors and exclude blue – with the exception of one unique molecule that, as it happens, can also attach to wool. Understanding the nature of blue, of color and color perception in general, and why the indigo molecule can defy all odds and yield a colorfast and permanent blue dye requires at least a general acquaintance with some concepts of physics, biology, and neurology.*

From the time of Aristotle, the dominant notion held that color represents an intrinsic property of objects – a notion still believed by many. Lipstick is red, we think, because it is made of red stuff. In reality, however, color is a particular effect of light, as you can see with a simple parlor trick: Shine blue light on a "red" object, and it will look black; shine yellow light on a "blue" object, and it now appears green. It's tempting to say that the yellow and blue lights merely distort the authentic blue and red colors, but, as the quotation marks indicate, those aren't the "true" colors at all. The colors we see result from ordinary light bouncing off objects and into our eyes. Since ordinary light is the light by which we see the world around us, we think that it reflects objects as they are. But it doesn't always.

We owe our modern understanding of light and color to the multifaceted genius of a seventeenth-century Englishman, Isaac Newton. England saw its last major outbreak of bubonic plague in 1665–66. As a precautionary measure, the students at Cambridge University, including young Newton, were sent home till

---

* It sounds more painful and less interesting than it will be.

the following year. In that year, the poet John Dryden wrote *Annus Mirabilis*, "year of miracles," commemorating that England had survived not only the Great Plague but also the Great Fire of London in 1666 (which some say helped put an end to the plague). The phrase was later applied to that miraculously productive year in the life and career of Newton. He spent his enforced vacation working on ideas underlying his later spectacular accomplishments: gravitation, laws of motion, calculus, and optics. Alexander Pope captured these achievements brilliantly in a famous couplet intended for an epitaph:

> *Nature and Nature's laws lay hid in night;*
> *God said, "Let Newton be!" and all was light.*[139]

One of Newton's most famous experiments involved the use of prisms. By directing a narrow beam of sunlight onto one side of a prism in a darkened room, he saw the colors of the rainbow projected onto a nearby wall. Before Newton, people had observed rainbow colors in prisms but believed those colors to be contained within the prisms themselves, as indeed they seem to be in ordinary unfocused light. Newton showed that color represented not a property of the prism but of the light itself, and that sunlight, also termed white light, actually contained the entire spectrum of colors. The tight beam of sunlight focused onto a prism split the light into its constituent colors, refracting or bending the different colored beams at slightly different angles, red on one side bending the least and blue on the other the most. Newton had demonstrated that the properties of light and color could be measured and quantified on the basis of what he termed "refrangibility," or refraction – that is, how much the prism bent each color.

Newton then went one step further. By shining the colored beams onto a second prism, he reversed the process so that the colors reintegrated into white. Further experiments revealed that selectively filtering colors and recombining them produced not only white, but different colors as well. On a stage, red, blue, and green spotlights, focusing together, form a white circle. Overlapping red and green beams appear yellow to our brains, indistinguishable from plain yellow, though structurally quite different. Newton himself created the notion of the color wheel, where the familiar spectrum of the rainbow – which schoolchildren know as a man named ROY G. BIV (red, orange, yellow, green, blue, indigo, violet) – fans out in a circular arrangement.

When white light, which contains all colors, is filtered and any one color is removed, our brains perceive the resulting mixture as the complementary color on the wheel opposite the absent color. In other words, when you take away blue, you see red orange, and the converse is true as well – take away red orange, and the resulting color will be blue. What we see as green results from white light losing both red and blue components. In general, the more of the spectrum removed and the tighter the remaining band of color, the more vivid the hue. Suppressing a small portion of the red-orange region leads to a muted blue, whereas filtering out almost all the colors except for blue will lead to the striking, crisp colors found in peacock feathers and lapis lazuli. Painters often refer to these different types of hues by the terms *saturated* or *spectrum*. Spectrum red is paint where only the blue has been removed, as opposed to saturated red, where only the red is present.

Pure or saturated colors can be measured by a quantity known as wavelength. Light, a form of energy emanating from the sun as electromagnetic radiation, reaches our atmosphere in broad waves. We describe waves, in terms of their length, as the distance from one crest or trough to the next. Their rate of oscillation, or frequency of undulation, correlates with their energy level. Just as with waves of water, slow undulations lead to long wavelengths and low energy, while rapid oscillations of greater frequency mean short wavelengths and higher energies. The electromagnetic spectrum contains "light" of all kinds, from low-frequency radio waves with wavelengths measuring hundreds of miles to high-energy gamma rays with wavelengths measuring less than the size of an atom. Those invisible waves make up most of the electromagnetic spectrum; our eyes can detect only a very small portion of it in the ROY G. BIV colors of the rainbow.

The range of visible light, with its myriad shades and color combinations, consists of wavelengths measuring from around 400 nm to around 700 nm (a nanometer being one billionth of one meter). The red portion of the spectrum incorporates the longer wavelengths of lower energy, but wavelengths longer than those fall into the invisible infrared area. As the spectral colors change through yellow to green, wavelengths get progressively smaller until they reach the shortest, most energetic lengths in the blue and violet region. Energies higher than those, corresponding to still shorter wavelengths, lie in the ultraviolet range that we can no longer see. The clear blue sky corresponds to a wavelength of between 474 and 476 nm.

Light waves form the first part of visual perception. Next comes the eye. According to the old notion of vision, our eyes send out rays or eye beams that enable us to see. The reality is almost the opposite. When open, our eyes encounter the stream of images composed of light that constantly bounces off our multicolored environment. That stream enters our eyes and hits the retina, the region of the eye densely packed with light-sensitive cells called photoreceptors. These are the famous rods and cones. The eye contains between 100 and 120 million rods, which, extremely sensitive to light, allow us to see at night, and about four to six million cones, which enable color vision. Cones come in three varieties, conveniently labeled L, M, and S for long, medium, and short wavelengths. L cones are particularly sensitive to red, M to green, and S to blue. Actually, all cone cells display a sensitivity in some degree to all wavelengths, but each type "specializes" in a region of color. Together, they enable us to see the entire range of spectral hues.[140]

Among mammals, this trichromatic vision is unique to humans and closely related species of monkeys and apes. Some insects also have three cones, but honeybees, for instance, can detect not only blue and green but also ultraviolet, which humans cannot see. Among birds, many have four cones, one sensitive to ultraviolet light and three in the visible region. Most mammals have only two cones, probably because they hunt nocturnally, color vision mattering less than highly effective night vision. Owls and bats, for example, have no cones at all – only rods.[141]

As all the light waves from all the shades of color of all the objects we see enter our eyes, they stimulate the cone cells, which in turn trigger other cells that send electrical impulses through the optic nerve to our brains. The whole system is enormously intricate and precisely calibrated to provide extremely fine color discrimination almost instantaneously. As we constantly move our eyes and absorb new images, millions of photoreceptor cells and different types of neurons cooperate in an amazingly efficient system of visual perception. The neurological activities of the brain, the biological features of the eye, and the physical properties of light waves combine to enable us to make visual sense of the world in which we live.

The first and most ubiquitous physical process that leads to the creation of blue is what gives the sky its magnificent azure color. For millennia philosophers and artists from Aristotle to Leonardo da Vinci speculated as to why the sky is blue. English physicist John William Strutt, third Baron Rayleigh, finally explained the true mechanism in the late nineteenth century. An extremely prolific scientist, Rayleigh made fundamental contributions to the fields of statistical and fluid mechanics, acoustics, and optics, and received the Nobel Prize in physics for his part in the discovery of the element argon. In 1871 Rayleigh married Evelyn Balfour, sister of Arthur Balfour, who, during his term as foreign secretary, authored the famous declaration in favor of a Jewish homeland in Palestine.

When light from the sun strikes air molecules and dust particles in the atmosphere, it sets them vibrating electronically, essentially turning them into tiny antennas that scatter the light in all directions in a process known now as Rayleigh scattering. Rayleigh's equations show the process to be more pronounced at shorter wavelengths, with blue light scattered nearly ten times more efficiently than red. When we look at a clear noon sky, we see the sun's blue light scattered back to our eyes by the atmosphere. When the sun sits on the horizon, we see the rays through an atmosphere that has scattered out the blue, leaving the beautiful reddish skies of dusk or dawn. A similar scattering process is responsible for the cerulean blue of a baby's eyes. Scattering also explains why clouds are white. Larger particles in the air, like water droplets in a cloud, scatter incoming light, but without preferring one wavelength over another. As a result, all the wavelengths of the spectrum scatter together, and we perceive the combination of colors as white.

A second method of producing blue has to do with a property of waves known as interference. When two stones drop into water, they send out ripples that soon merge into each other and create patterns as the concentric wave fronts cross one another. At any point, such interference can be constructive or destructive. When waveforms align crest to crest and trough to trough, they amplify each other constructively. Out of sync, they cancel each other out. When light reflects from very structured and regular patterns, the geometry can favor specific wavelengths by inducing constructive interference of a specific color. The brilliant hue of the blue morpho butterfly and the bright blue of some flowers, for example, reflect this type of structural color generation. Sometimes the color depends on the angle of reflection, as when we see a rainbow effect on a CD with its regular pattern of micro grooves, or on a soap bubble or a gasoline spill where light reflecting off the inner and outer surfaces generates interference effects. That reflection angle also causes the metallic sheen of beetle wings and the iridescent plumage of a peacock.

Complicated as the interaction among matter, light, and our brains may be, it's nothing compared to the complexities that modern quantum physics introduced to the subject. When Newton first split light into its component colors, he believed that light consisted of "corpuscles," or (in modern terms) particles. In the early 1800s, however, another British genius, Thomas Young, convincingly demonstrated the wave nature of light.

Young had mastered at least a dozen languages, including Arabic, Hebrew, Persian, and Turkish, and had worked on deciphering the hieroglyphics of the Rosetta stone before Champollion. His famous double-slit experiment shone light through two parallel slits and recorded the results on a screen behind the slits. A pattern of dark and light bands fell onto the screen, which meant only one thing: The pattern must have resulted from interference. The light and dark bands that he saw represented light moving through the slits – not as if consisting of particles moving in a neat straight line but rippling outward – the ripples or waves in alignment in some places and out of sync in others.

Despite Newton's enormous prestige, Young's elegant experiment became the defining view for the nineteenth century. But in 1905, another scientific annus mirabilis, Einstein came along and revolutionized physics with four papers, including one on the photoelectric effect. It showed that light behaves not only as a wave but also as a particle, giving rise to the concept of wave-particle duality. The notion was baffling. German physicist Max Planck described the unit of light as a wave packet; Einstein called it a light quantum. Today we call it a photon.

Concepts such as wave-particle duality run counter to our normal thinking and experience, but in the weird quantum world of arcane equations solving formidable problems related to un-observable phenomena, they are quite possible. Young's double-slit experiment had proved that light behaves as a wave, and Ein-stein's photoelectric effect proved that light behaves as a particle. As physicist Richard Feynman, one of the main contributors to the quantum theory of light, wryly stated: "I can safely say that nobody understands quantum mechanics." Yet, even without un-derstanding it, physicists have successfully used its concepts and formulas to explain a wide variety of phenomena and to contrib-ute tangible benefits to our everyday world.

Quantum mechanics also accounts for the relative scarcity of the color blue in our physical environment. Newton showed that filtering out or removing certain colors from the visible light spectrum resulted in the perception of other colors. Filtering red leads to blue; removing blue gives red; filtering both red and blue gives us green. In molecules, this process, called absorption, is the main source of all color generation. When hit by a photon, a molecule can absorb its energy, and each molecule has a unique set of specific wavelengths that it will absorb.[142] Blue is so scarce because not many substances absorb light in the red sphere. Since creating blue requires absorbing red, nature is constrained from generating that color. The blue of the sky presents something of an anomaly because it results from the scattering of blue wave-lengths and not the absorption of red ones.

The color of the ocean represents the third way in which nature creates blue; it is not, as people sometimes believe, a re-flection of the blue sky. Water, for complex reasons only partly understood, just barely manages to absorb some red, so that it appears blue. Since this absorption is very weak, you need huge volumes of water in order to perceive the color. That's why a cup of water looks clear, a swimming pool has a slight blue tinge, and the vast ocean appears deep blue.[143] Aside from the ocean, however, the process of absorption, which accounts for the colors of virtually all the objects in our world, does not yield blue.

The fourth way that nature creates blue involves a process described by ligand field theory. Atoms in a crystal, called ligands – from the Latin *ligare,* "to bind" (whence the word *ligaments,* which bind our bones together) – surround a central atom and bond with it. When a metal impurity enters a crystal lattice, it can distort the electric field and create a situation where electrons pass from one atom to another. This charge transfer, or electronic transition, often happens at energies within the visible range, producing the striking colors of gemstones. Aluminum oxide, for example, forms the colorless crystal known as corundum, or colorless sapphire. When chromium – from the Greek *chroma* (χρῶμα), meaning "color" – replaces a few of the aluminum atoms, rubies form. Blue sapphire forms when titanium and iron atoms contaminate the corundum crystals.

The regular structure of crystal gems forms at high temperatures and intense pressures, such as those found in the Earth's interior. Biological processes cannot create them, so charge transfers don't happen in organic substances. Blue gemstones such as aquamarine and rocks of lapis lazuli get their colors from crystal ligand processes. The intense color of lapis lazuli, already being quarried more than five thousand years ago, made it a favorite source of jewelry all over the ancient world. The Prussian blue crystals that the Rebbe of Radzyn created from cuttlefish ink mixed with iron resulted from the same inorganic ligand process and required the intense heat of the Hasidic cauldrons.

The ancients didn't have access to the knowledge or technology required to create our synthetic blue dyes; instead, they sought natural substances to produce that color. But making blue naturally, it turns out, is not a simple matter. The absorption of a photon can excite a molecule, causing it to rotate or vibrate at a higher energy level. Atoms and molecules can also channel absorbed energy by rearranging the structure of the electron clouds that surround them and form the bonds between them.

When first learning about them, we often picture electrons as tiny planets in a miniature solar system cleanly orbiting a nucleus. The equations of quantum mechanics, however, give us a far different view of electrons: diffuse clouds of exotic shapes resembling clovers and teardrops stemming from the nucleus. Atoms can change their energy states when electrons shift from one cloud pattern to another. According to quantum theory, the wavelengths required for a photon to excite these electron transitions typically measure shorter than those of visible light and fall into the ultraviolet region. But exceptions occur. The many patterns that form around and between atoms sometimes allow for transitions at lower energy levels, corresponding to wavelengths acting in the visible range. Those wavelengths, however, still tend toward the higher energy side of the visible spectrum, namely the blue and violet region. Zooming back out, when objects absorb these photons, the colors that result look red – which explains why our world consists primarily of earth tones.[144]

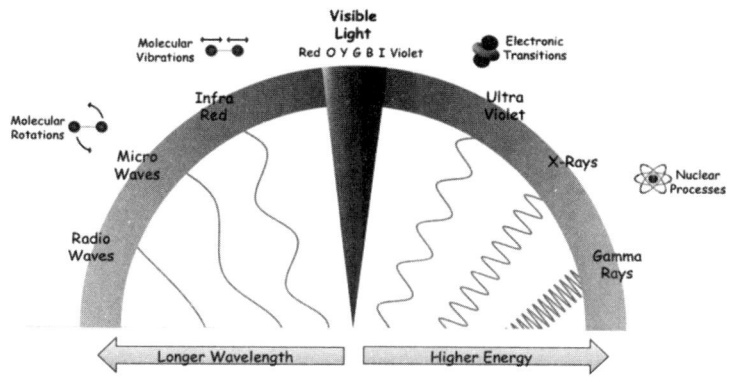

Electromagnetic spectrum

In order to achieve blue by means of the fifth method, a substance must absorb photons only in the red-orange range. But the physical processes of energy transition in atoms and molecules do not correspond to photons of that wavelength. Rotational and vibrational transitions are caused by invisible infrared photons, while electronic transitions are generated by blue and ultraviolet photons. Blue formed by the absorption of red-orange photons occurs extremely rarely since the atoms and molecules that constitute matter simply have no way of using those photons.

The exception is a lovely little molecule – produced by some plants in the pea family and also by some sea snails – called indigo.

The laws of nature seem to have a predilection for beauty, elegance, and simplicity; in physics the yardstick that measures those qualities is symmetry. In nature greater symmetry

The indigo molecule

generally leads to greater stability. One possible explanation of the unique characteristics of light absorption by the indigo molecule is based on this principle. Though a rigorous understanding of how indigo interacts with light requires complex quantum mechanical modeling that taxes the strength of even the fastest computers, on a more basic level, symmetry that we can comprehend by intuition will suffice.

Rotating an indigo molecule 180 degrees leaves the structure unchanged. In other words, if you draw the molecule on a sheet of paper and then turn the paper upside down, all the atoms are in the exact position as they were before the rotation. When an indigo molecule is excited by absorbing a photon, the electrons redistribute themselves somewhat, the bonds differ slightly, and some of the atoms carry a charge. But this new configuration of the excited molecule retains the same properties of symmetry as the molecule in its original ground state; it, too, remains identical when undergoing a 180-degree rotation.

Indigo is a compact and simple molecule that exhibits a high degree of symmetry in both its ground and excited states. As a result of the convergence of these characteristics, shifting between states requires relatively little energy compared to other

Excited state of the indigo molecule
COURTESY OF R. M. CHRISTIE

molecular transitions. These lower energy requirements correspond to photons of longer wavelengths, which accounts for the unique property of indigo to absorb in the red-orange region of the spectrum. The peak absorption centers at 613 nm, which results in our perception of a beautiful azure blue.[145] Add a bromine atom or two, and the elegant simplicity lessens so that the absorption peak moves to slightly shorter wavelengths. This change pushes the color of mono- or dibromoindigo away from pure sky blue toward purple or burgundy.[146]

In the natural world, blue objects occur quite infrequently because the physical processes that lead to that color are few and relatively uncommon. Organically blue materials are even rarer. The conditions required mean that creating a natural blue dye is an almost impossible feat – still a daunting task even with the tools available to modern technology. The solution that nature has provided in the form of the indigo molecule proves nothing short of amazing.

It truly is the rarest blue.

XIII
RHAPSODY IN BLUE

$\mathscr{C}$olor defines the state of the natural world. It characterizes every object we perceive. We differentiate one object from another not by sharp lines, as in drawings, but by changes in contiguous colors. Color perception is an extraordinarily complex physiological process, and a decidedly psychological component complicates it even more.

The introduction, proliferation, and popularity of man-made dyes such as blue, purple, and red in ancient times testifies to a psychological need, a hunger for color and richness. The Bible's description of the extravagant garden courtyard of Ahasuerus, king of Persia – "there were hangings of white cotton and blue wool, caught up by cords of fine linen and purple wool to silver rods and alabaster columns; and there were couches of gold and silver on a pavement of marble, alabaster, mother-of-pearl, and mosaics" – underscores color as a main feature of opulence and luxury.[147] The great sculptures of classical Greece didn't originally have the cool polished look we see today; colored paints covered them, faint traces of which remain. But not every ancient culture cherished sumptuous colors.

In many societies, writes historian Robert Finlay,

the elite displayed pronounced chromophobia. In their eyes, bright color was the province of barbarians, children, the rabble, and the ignorant; it exemplified the superficial, subjective, irrational, self-indulgent, sensual, disorderly, and deceptive. As an integral part of this sweeping defamation, color was despised inasmuch as it was seen to be cherished by women, who were (all men agreed) vain, flighty, foolish, shallow, seductive, and driven by emotion.[148]

Color belonged to the unsophisticated and was the province of the lower classes. As Goethe, observing this bias in modern times, put it: "People of refinement have a disinclination to colours."[149]

The reason for the chromophilia of some cultures such as Persia and the chromophobia of others, Finlay suggests, has more to do with geography than with matters of taste. The two major Near Eastern cultures developed primarily along narrow slices of land on either side of great rivers, the Nile in Egypt, and the Tigris and Euphrates in Mesopotamia, which literally means "between the rivers." The dreary landscape of the arid, inhospitable, sand swept deserts on either side of those thin strips presented a monotonous beige devoid of any bright tones. Life existed along the water among an abundance of the vibrant colors of grasses, trees, and flowers. For the people of the Fertile Crescent, color came to represent life and civilization.

Ancient Chinese and Japanese cultures, in contrast, displayed a definite prejudice against color. In the form of a major Chinese style of painting known as Shui Mo, or "water and ink," monochrome ink painting represented an ideal, the goal in painting an object being to "seek its supreme unborn nature beyond all form and color." [150] A superficial and trivial distraction, color interfered with the true essence of a form. It was specifically in the absence of color that an object's true nature could be found.

Finlay suggests that this prejudice against color may have roots in early religious rituals. The priests of the Shang dynasty (second millennium BCE) sought to read messages from the gods by heating animal bones or turtle shells and examining the resulting fracture patterns.

Oracle bone inscription from the Shang dynasty (second millennium BCE)
© JUN MU | DREAMSTIME.COM

The linear designs on these "oracle bones," considered of heavenly origin, had a profound effect on the development of Chinese culture, their style heavily influencing the development of the Chinese pictographic writing system. Another and more subtle consequence of divine communication coming via these cracks and lines on bones was that black and white came to be considered the paradigm for visual representation.

If the absence of color has aesthetic and psychological implications and if colors themselves also satisfy aesthetic and psychological needs, we have entered the realm of ambiguity and paradox. That realm particularly appeals to a leading twentieth-century Jewish religious scholar and philosopher. Scion of a family of great, innovative rabbinical sages, Rabbi Joseph

Soloveitchik, a preeminent spokesperson for modern Jewish thought, brought a unique mixture of philosophy and penetrating psychological insight to his study of Bible and Jewish law. Aware as he always was of apparent contradictions, and attuned to seeming absurdities, he often found symbolic interpretations to reconcile differences or to provide profound insights into traditions, laws, and life.

Among the topics that engaged him was the wearing of tzitzit and the requirement of a blue thread. Such a requirement might seem picayune, arbitrary, irrational. What possible significance could a blue thread have? If blue was so significant, why not all blue rather than a mix of blue and mainly white? In his explanation of the symbolism underlying the command to wear white and *tekhelet* blue strings on the corner tzitzit, Rabbi Soloveitchik brings to light a profound existential reality that touches on a core element of the human condition.

Rabbi Soloveitchik maintains, as Finlay had suggested, that the color white symbolizes that which is clear and understood, that which the rational mind can comprehend and the intellect can grasp. Conversely, the deep blue of the sky and ocean signifies what cannot be grasped, what is mysterious and enigmatic, all that lies beyond our cognitive abilities, beyond the horizon of human intellect and experience, in the murky depths that we cannot fathom. White represents what we know, what we can measure directly and tangibly, what we can control, and what we can predict. Blue is intangible, everything we sense and feel but cannot fully substantiate. It lies beyond our control, taking us by surprise and changing our destiny.

Modern man strives to make the world as clear and comprehensible as possible; the goals of science and technology include a full understanding of the deepest secrets of nature, quantifying its parameters, taming its savagery, and harnessing its power. But nature does not always oblige; she confounds and frustrates those who seek to impose order and regularity. This holds true from the greatest level, where chaos reigns in the stellar explosions of supernovae, to the smallest, where a single microscopic virus or rogue cell can wreak havoc in the body and abruptly end a life.

Each of us, during the course of life, experiences both white and blue. At times, everything proceeds according to plan, and we feel that we have it all figured out and are in control. At other moments, we find ourselves struggling with uncertainty and doubt, unsure of what course to take, plagued with despair. What is true of an individual is true of relationships as well. Even the most loving couples and the closest families go through "blue" periods of tension, anxiety, and friction. Human life consists of a constant ongoing dialectic.

This dichotomy also plays out in historical processes and world events. At times, the world makes sense, and we confidently formulate theories that embody our experiences and expectations. World leaders believe that interactions between nations and cultures can follow defined rules and logical, agreed upon principles. But even the greatest strategists and statesmen encounter intractable forces that compel them to abandon their best-laid plans. Often they find themselves led by events, not leading them. The blue thread on the tzitzit doesn't only remind us, as the Bible puts it, of an obligation to obey God's commands. Affixing both blue and white strings to our clothing, says Rabbi Soloveitchik, reminds us of our precarious place in the world and the need for faith in the face of uncertainty.[151]

RHAPSODY IN BLUE

The tension between the rational and irrational that defines the human condition offers one powerful symbolic reading of the color blue. Symbolic interpretations, however, tend to multiply, and blue has taken on a bewildering variety of other meanings. It is the color of desire and of yearning for the infinite, for what will remain forever out of reach. The blue flower (*Blaue Blume*) that represents this yearning in a famous novel by Novalis became a defining symbol for the Romantic movement in Germany.

On the other hand, blue represents the color not of doubt or the intangible and mysterious, but of steadfastness, trustworthiness, and fidelity – true blue. In Islamic cultures blue protects, and blue doors ward off evil spirits. In medieval paintings the Virgin Mary often wears a blue robe, a custom originating in the early Byzantine Empire when it was the clothing of an empress. The Hindu god Vishnu and his avatars Ram and Krishna are blue; indeed, the name *Krishna* in Sanskrit means "blue" or "black." In the kabala, the Jewish mystical tradition, blue is often associated with the lowest of the ten *sefirot* (emanations) – *malkhut* (kingship) – the bridge between the physical and the spiritual realms. Some suggest that this has to do with the fact that blue inherently represents opposites, both sea and sky, both fire (as in the center of a candle's flame) and water.[152]

Frequently, however, objects are blue for no clearly discernible reason. Masonic blue is the supreme color of Freemasons, sometimes explained with reference to the Biblical blue of Temple accoutrements – but that's not entirely convincing. Why are blue ribbons awarded at state fairs to winners who raised the fattest chicken or baked the best blueberry pie? Perhaps those blue ribbons derive from the English blue ribband, the prize awarded for over a century to the ships that broke records crossing the Atlantic, but why was that blue? It may have evoked the Most Noble Order of the Garter, with its blue emblem – but at a certain point color symbolism becomes arbitrary or inconsistent.

The color blue also has negative connotations, of course, associated with feelings of depression and despondency. The phrases "feeling blue," "singing the blues," and, particularly descriptive, "having a fit of the blue devils," all common, imply melancholy, although the origin of the association of that feeling with the color remains unclear. The U.S. Navy's website claims the phrase "feeling blue" traces to the old practice among sailing ships of flying a blue flag when returning home as a way of announcing that men had died during the voyage.

More solidly founded are the psychological effects of color and the specific emotional reactions they elicit. These reactions are not arbitrary nor always culturally determined; people from diverse backgrounds and societies tend to share emotions associated with different colors. In his comprehensive *Blue: The History of a Color,* Michel Pastoureau contrasts red, which excites, with blue, a "calming" color. It's no coincidence that blue – perceived by most people as peaceful, neutral, unthreatening – is the color of the flags of the United Nations and of the European Union. Studies have shown that people working in a blue room can solve twice as many insight puzzles as in a red room since blue allows for more reflective and associative thinking. The notion that blue psychologically can calm and relax people underscored the 2006 decision by the Japanese transit authority to install bright blue

LED lighting in subway stations in an attempt to reduce a recent spate of suicides. Web pages advocate blue-light therapy in which the patient undergoes exposure to a light box and is bathed in blue light, in order to regulate the production of melatonin, a hormone that plays an important part in inducing sleep."[153]

Beyond psychology and into the realm of pharmacology, the late Oliver Sacks recounts his hallucinogenic quest to find blue. He describes the incredible trip in a New Yorker article:

I had long wanted to see "true" indigo, and thought that drugs might be the way to do this. So one sunny Saturday in 1964 I developed a pharmacologic launchpad consisting of a base of amphetamine (for general arousal), LSD (for hallucinogenic intensity), and a touch of cannabis (for a little added delirium). About twenty minutes after taking this, I faced a white wall and exclaimed, "I want to see indigo now — now!"

And then, as if thrown by a giant paintbrush, there appeared a huge, trembling, pear-shaped blob of the purest indigo. Luminous, numinous, it filled me with rapture: it was the color of heaven, the color, I thought, that Giotto spent a lifetime trying to get but never achieved — never achieved, perhaps, because the color of heaven is not to be seen on earth. But it existed once, I thought — it was the color of the Paleozoic sea, the color the ocean used to be. I leaned toward it in a sort of ecstasy. And then it suddenly disappeared, leaving me with an overwhelming sense of loss and sadness that it had been snatched away. But I consoled myself: yes, indigo exists, and it can be conjured up in the brain.[154]*

---

* In a personal correspondence, I noted that I too have spent my life searching for indigo (minus the pharmacological launchpad)

Blue has long served as a wellspring for artistic creativity, one of the most famous examples being Picasso's Blue Period, a reaction to the suicide of his best friend, Carlos Casagemas. One of the most extreme examples of blue as inspiration exists in the work of Yves Klein. He created a specific shade of blue that he patented and became known as International Klein Blue. Even in earlier work he focused on blue, but eventually he created only monochromatic blue works, often painting nude models blue and having them slither and slide on a canvas. He summarized his passion for blue in a remembered experience: "As I lay stretched upon the beach at Nice, I began to feel hatred for the birds which flew back and forth across my blue sky, cloudless sky, because they tried to bore holes in my greatest and most beautiful work."[155]

Perhaps the most profound insight comes from the great Russian artist, Wassily Kandinsky. He had a fascinating neurological condition known as synesthesia where he experienced color both visually and audially. Each color had its own tone, and for him vision had a musical component. Kandinsky speaks of "the inclination of blue towards depth." He writes in *On the Spiritual in Art* (1911-12):

"The deeper the blue, the more it beckons man into the infinite, arousing a longing for purity and the supersensuous. It is the color of the heavens just as we imagine it, when we hear the word heaven.[156]

---

and described our work with *tekhelet*. Dr. Sacks resonded, "I do not know whether the blue of my tallit, as a boy, this mythical/mystical blue, played any part in determining my hallucination years later. I wonder if I will ever see it again."

The blue of the *tekhelet* strings can lend an additional dimension to melodic inspiration. The late composer and pianist Moshe Cotel served as professor of music composition at the Peabody Conservatory of Johns Hopkins University. A musical prodigy, he had studied there since the age of nine, wrote a full-scale symphony at thirteen, and won the American Academy of Rome Prize for music composition when he was twenty-three. In later years he moved closer to his Jewish roots and received ordination as a rabbi, using his music as a medium to reach his congregants. Blue *tekhelet* threads inspired *Chronicles: A Jewish Life and the Classical Piano,* one of his last compositions. Cotel describes the blue fringes on the prayer shawl and the mysterious mollusk source of the color, ending the work with an interpretation of Gershwin's *Rhapsody in Blue.*[157]

Inge Boesken Kanold developed an interest in ancient colors during her fourteen-year trip to the Far East. After working with Javanese artists and writing a manual in Indonesian on the preparation of canvas for use in tropical climates, she left for Beirut, Lebanon, fascinated by the purple shellfish dye that was once the pride of that region. Her subsequent work in Germany and France pushed the envelope in terms of the use of shellfish dye in art. She discovered a method for dyeing parchment purple; reinvented the ancient process of painting with the dye, known in antiquity as *purpurissum;* and discovered a difference in color nuance between male and female snails. She often creates her magnificent artwork by pressing the snail glands directly against the canvas. In her own words, "working with *tekhelet* is the spiritual summit of my artistic life."[158]

Artists find inspiration in color and express their innermost creative self through it. Philosophers, however, have questioned whether color has any independent reality at all outside the mind. "I cannot believe," Galileo said, "that there exists in external bodies anything, other than their size, shape, or motion."[159] In *The Objective Eye: Color, Form, and Reality in the Theory of Art*, John Hyman takes up Galileo's opinion, which, Hyman writes,

> *is still widely held today. For example, John Gage, who has made the place of color in art the main theme of his work, claims that "Newton… showed that colour was indeed illusory." Many student textbooks on visual perception treat this claim as an established fact. For example, Stephen Palmer states that "colour is a psychological property of our visual experiences when we look at objects and lights, not a physical property of those objects and lights." And similar claims appear in many learned articles as well. For example, in an important article on color vision in monkeys, Samir Zeki claims that color "is a property of the brain, not of the world outside."[160]*

Assuming that color belongs to the observer's mind rather than the object's existence, color perception, far from being objective and universal, should be completely subjective, varying from person to person and also from culture to culture.

In a fascinating 1969 experiment, Brent Berlin and Paul Kay at the University of California at Berkeley set out to prove a related point, that color language is relative, and that no semantic universals exist to describe how we see the world. They found the exact opposite to be true. Studying languages from around the world, they discovered that no language ever contained more than eleven basic color terms. Furthermore, color words appear in a predictable sequence: Languages with only two color words have black and white. Languages with three add red – not green or purple – and as more color words appear, the additions in order are yellow, green, and finally blue. Presumably blue falls last in the line to gain recognition because of its scarcity in the natural world.

The researchers next asked native speakers of twenty diverse languages to group 329 colors into their language's basic terms to see where, according to each system, one color ended and another began. Their results proved two universal truths about color perception. First, the center of a basic color term held across all languages, and, second, all subjects grouped colors comparably. That is, though obviously the word for "blue" is different in Swahili and Cantonese, a person living in Mozambique and someone in southern China both identify the range of colors termed "blue" and the center of that range as the same. This seminal work proved that color perception is hardwired in the brain, universal across cultures and ethnic groups, paving the way for subsequent researchers such as Oliver Sacks and Steven Pinker to demonstrate that so many of our behaviors, aptitudes, and conceptions indeed have a neurological basis.

The argument that color perception is universal and cross-cultural should lead us to believe that it crosses time boundaries as well. William Gladstone served as British prime minister four times and also found time to write a three-volume study of the works of Homer. One chapter of his *Studies on Homer and the Homeric Age* deals with the perception of color in the Homeric tradition. Gladstone claimed that blue doesn't exist in Homer's vivid language, even in descriptions of objects that we would assume call for that color. The sea is "wine-dark," black, gray, even purple – but never blue. Nor is the beautiful azure Mediterranean sky described as blue. To account for this peculiar lacuna, Gladstone proposed that the ancient Greeks were blue blind, physiologically unable to perceive the color blue. They simply did not see the sea and the sky as we do now. His conclusion shocked and disturbed people who saw in classical antiquity the height of human achievement.

Promoted and derided for the past 150 years, Gladstone's radical thesis reared its head again most recently in Guy Deutscher's *Through the Language Glass,* where he tried to get to the bottom of the controversy by conducting a language experiment on his daughter. As she learned to talk and identify colors, he made sure never to speak to her about the blue sky. At eighteen months, when she could easily recognize and label blue objects, he took her outside and asked her what color the sky was. At first, she couldn't answer, but eventually she said it was white. It took her another few months before she consistently called the sky blue. Deutscher concluded that neither the child nor the ancient Greeks were unable to see blue or label objects of that color. Rather, it was a question of focus.

How an object appears depends upon a number of different components that make up our perception, and we lump many of those together under the general term *color.* The language of the ancient Greeks concentrated more on luminosity than on hue. They grouped dark shades or light shades together, so that the dark sea could be black or wine colored. We, on the other hand, do focus on hue, grouping both light and dark shades of the same color together under the same term, using the word *blue* both for the color of the sky at midday and evening.[161]

Look up at the boundless sky or gaze at the endless sea, and you might find yourself exclaiming with Nabokov – or at least appreciating his sentiment – "What bliss there is in blueness. I never knew how blue blueness could be."[162] Whatever prompted Nabokov's exclamation would have appeared as exactly the same thing to any of us: Light waves hitting the retina, exciting neurons to fire, and sending impulses to the visual cortex produce the same phenomenon of vision. But Nabokov clearly had something else in mind besides visual cortex activity.

His very personal, subjective reaction resulted from a visual episode filtered through his life experiences, beliefs, emotional history – his whole personality. Numerous psychological studies have demonstrated that what people believe can transform their perception of reality and that both our attitudes and mindset definitively affect not only our assessment of an object or experience, but also our emotional and physical responses to that object. In other words, we can choose to see the serene bliss of blue or its deep melancholy, or we can choose to see mere blue. For the most part colorless, clear, and transparent, air and water lack any inherently inspiring qualities or beauty, but they also give rise to the intense blue that permeates our world. In them we can see nothing, or we can see the magnificent azure of the gleaming sky or the deep blue of the shimmering ocean.

XIV
TANGLED UP IN BLUE

A fine line exists between stubbornness and persistence, between dedication and monomania. Which is the right term to describe the visionary, the dreamer ready to travel centuries into the past, armed only with a few fragmentary facts, with vague, confused, contradictory information, in search of a mythical creature from which, by some half-known process, it might be possible to extract the semblance of an ancient dye? With the erstwhile existence of sky-blue threads its only certainty, the entire enterprise seemed bound to fail, its chances of success almost nil.

Dissatisfied with his world and the course of his life, Don Quixote set out to change reality. Reality resisted change and ultimately defeated him. But not every quixotic quest ends in failure. We have seen the true *hillazon* rediscovered, the technique of producing dye from its gland reinvented, and a biblical practice dormant for centuries reestablished.

Once Otto Elsner discovered the effects of sunlight on the murex dye solution and solved the riddle of *tekhelet*'s sky-blue color, Jews could again wear authentic blue threads on their tzitzit.

News of the discovery, however, reached the religious community slowly, many of them shunning secular learning on any level and especially suspicious of scientific knowledge. Even among those with a less negative disposition toward science, not many had access to the *Proceedings of the Seventh International Symposium on Fiber Science and Technology,* the conference having taken place in Hakone, Japan, in 1985. That's where the findings of Elsner and Spanier first appeared.

One exception, Irving Ziderman, a religiously observant Jew as well as a practicing chemist, wrote a number of articles explaining that authentic *tekhelet* had been found and urging the Jewish community at large to take a serious interest in the new discovery. Rabbi Bezalel Naor also attempted to popularize the topic, as did Rabbi Menachem Burshtein, who wrote a detailed examination of the works of the Radzyner Rebbe and Rabbi Herzog. Despite all these efforts, however, *tekhelet* research remained confined primarily to the laboratory and the library – until one article reached the desk of a young rabbinical student.

Eliyahu Tavger grew up in a house that cherished the ideals of freedom of thought, devotion to a cause, and sacrifice on behalf of others. His father, Ben-Tzion Tavger, a Russian physicist, had been persecuted by Soviet authorities for promoting the rights of Jews to study and practice their religion freely. He was dismissed from the University of Gorky, incarcerated in a mental institution, and exiled to Siberia. Eventually he emigrated to Israel along with his teenaged son. There the renowned professor became involved in the struggle to prevent desecration of Jewish sacred places and fought for the right of Jews to pray in holy sites such as the Cave of the Patriarchs. He even left his position at Tel Aviv University at one point to work as a watchman in the

Jewish cemetery in Hebron, where he carried out clandestine excavations by night and helped unearth the city's ancient synagogue, buried under rubble and trash for more than fifty years.

In his early twenties, Eliyahu Tavger began to investigate a manuscript of a ninth-century Babylonian rabbi on the laws of tzitzit. In the course of his studies he grew increasingly interested in *tekhelet,* which soon took over as the major focus of his research. Only a few singular individuals devote their lives to a subject, tireless in their efforts to probe the deepest levels of all relevant knowledge, to translate that knowledge into practice, and to bring ideas from potentiality to reality. The story of the rediscovery of *tekhelet* has featured the efforts of Gershon Henokh Leiner, the Rebbe of Radzyn, and Isaac Halevi Herzog, chief rabbi of Israel. Rabbi Eliyahu Tavger stands in their elite company.

Tavger's early attempts at collecting snails, fabricating dye, and dyeing wool met with frustration and disappointment. He had undergone the traditional rabbinical training in ultra-Orthodox schools renowned for their Talmudic excellence – but not exactly for eminence in the areas of marine biology and color chemistry. The few researchers who had succeeded in dyeing wool with murex extract had done so in modern, well-equipped laboratories and with ample funding. Tavger was carrying out his work on the beach, with old knives and broken buckets, on the meager stipend of a rabbinical student. The university professors to whom he turned for guidance were not disposed to offer help, besides which very few had any practical knowledge of the exact methods, ingredients, or proportions of the dye process.

Tavger concluded that in order to move ahead, he had to enlist the help of industry professionals, and he turned to a leading Israeli dye laboratory. The lab director told him and his father-in-law the terms of the contract. The institute would help develop the methods for dyeing with murex snails, and in return Tavger would grant them perpetual royalties on every string manufactured in addition to an up-front fee of five thousand dollars. As they left the meeting, his father-in-law remarked to Tavger how disappointed he was with the terms, especially the fee – an astronomical figure given their financial circumstances.

"This will give us the chance to fulfill one of God's commandments for the first time in 1,300 years," Rabbi Tavger replied. "I consider the price quite a bargain."

Resigned to accepting the terms, they decided nevertheless to think it over for a few days. That night, Tavger received a call from Danny Levy, a chemist friend helping him with the *tekhelet* project. Levy had just come across a paper from a recent Japanese congress in which the author described his experiments with murex dyeing. It was, of course, the Elsner and Spanier paper, and it described the process in enough detail to provide Tavger with the clues that he required.

In 1988 Rabbi Tavger finally succeeded in dyeing wool with glands taken from *Murex trunculus* snails. He took the blue wool to the foremost tzitzit manufacturer in Jerusalem, who produced a few strings from it. Eliyahu Tavger tied those strings to his prayer shawl and recited a blessing reserved for the most auspicious and momentous events in one's life: "Blessed are You, O Lord our God, King of the universe, who granted us life, and sustained us, and allowed us to reach this occasion."

THE RAREST BLUE

He and his father-in-law never signed the contract with the dye lab.

Joel Guberman – who had contacted English medievalist John Edmonds about woad fermentation – and I have been friends since our high school days. In 1988 a careless driver ran the curb and killed Joel's brother Avi as he worked in his yard. Devastated by the tragedy, Joel sought to commemorate his brother's life by committing himself to practice a biblical precept in a more meaningful and more thorough manner. The precept he chose was tzitzit.

In a Jerusalem library where he was researching the subject, a librarian remarked that another gentleman was doing similar research and pointed out Irving Ziderman. The two discussed the topic, and Ziderman told Joel of Rabbi Eliyahu Tavger's recent success in producing threads of genuine *tekhelet*. The major obstacle, he explained, was that collecting the snails was proving terribly difficult; if only they knew some scuba divers willing to help, the job could be done much more efficiently. If that was the problem, Joel assured him, he could take care of it.

Joel immediately called me and another close friend, Ari Greenspan, to ask if we would be interested in diving for snails. Our curiosity piqued and always up for an adventure, we immediately agreed.

A few days later, Joel, Ari, and I piled into Ziderman's old station wagon and headed north. We sat riveted the entire three-hour ride as Ziderman recounted the story of the loss and rediscovery of *tekhelet* in all its amazing detail. He introduced us to the

Radzyner Rebbe, Rabbi Herzog, Lacaze-Duthiers, Pliny the Elder, Otto Elsner. He told us about the *Murex trunculus,* about the
gland that produced the dye, and how esteemed *tekhelet* and Tyrian purple had been in ancient times. He showed us the snail itself
and cautioned us to scrutinize it carefully, explaining that underwater a camouflage coating was going to make it nearly
impossible to distinguish from surrounding rocks. Fascinated
and utterly absorbed, we could barely believe that, as Ziderman
stopped the car, the long drive had ended. We emerged to meet
a soft-spoken man, with gentle, sparkling eyes and a humble smile,
and shook hands with Rabbi Eliyahu Tavger.

At a dive shop in Acre's quaint old city district, we picked up
gear and headed down to the ancient port. Decked in full scuba
equipment, Ari and I flippered our way into the sea. The waves,
choppy and strong, kept knocking us over. When we finally managed to get underwater, we found that the wave action had
churned up the bottom sand, making for very poor visibility. We
had to give up, disappointed and annoyed that we had let the
group down. Rabbi Tavger, however, suggested that we try another spot about an hour farther north, almost at the border with
Lebanon.

There we found a large natural lagoon formed by coral rock
and kurkar that protected us somewhat from the waves. In the
shallow water, our oxygen tanks lasted quite a long time, and over
the next two hours Ari and I collected 293 snails. Back on shore,
we watched as Rabbi Tavger broke open a shell and extracted a
gland. We joined the messy work. Completing the dyeing process
took a few more weeks, and when the time came, Rabbi Tavger
called us to join him as he prepared the dye solution, exposed it
to sunlight, and dipped the wool into the liquid dye. I will never

forget that first experience of seeing the dripping wool emerge a dirty brownish yellow that slowly changed to a magnificent sky blue.

An expert Jerusalem tzitzit maker later spun the wool into threads and plied the threads into strings that we tied to our prayer shawls. We learned that different leading rabbinical authorities recommend different ways of tying the strings. The Bible merely says to "attach a cord of blue to the fringe at each corner," but Jewish tradition didn't stop with that simple, somewhat vague instruction, later expanding upon the theme. Different styles of embellishing the threads developed, with intricate macramé stitches and fancy knots and twists creating eye-catching patterns with the white and blue threads. Some of these designs were chosen for their underlying religious symbolism and significance, others for their aesthetic appeal.[163]

Tying the strings for the first time, I felt an exuberant fervor – religious as well as intellectual – mixed with an overwhelming sense of humility born from contemplation of the devout Jews over the past millennium who only could have dreamed of holding in their hands the blue strings that from now on I would be able to look at every day.

Something else happened that day on the beach.

Joel, Ari, and I came to a shared decision; we would join with Eliyahu, dedicate time and effort to studying the topic of *tekhelet,* and make information – along with the actual blue strings themselves – available to the world. To tell the truth, we all became somewhat obsessed. Within a few months, we had founded a

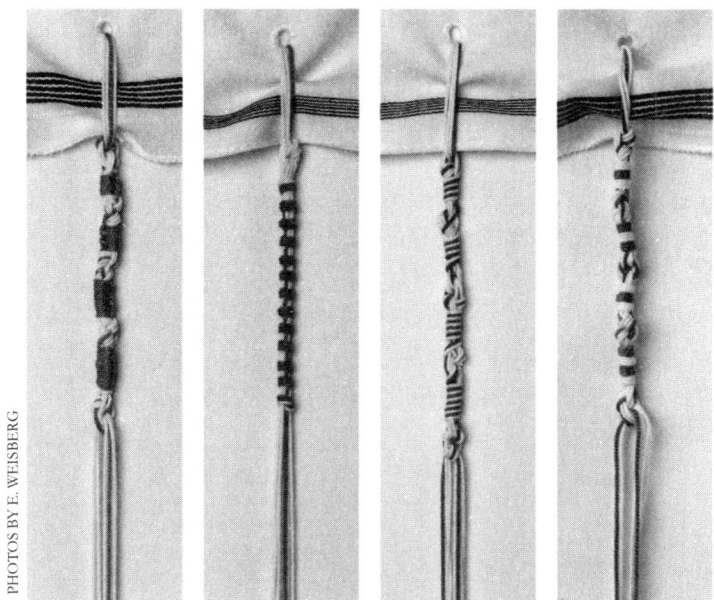

Samples of tzitzit with tekhelet *threads*

nonprofit organization, Ptil Tekhelet, and, backed by the absolute
support and encouragement of wives and family, we began to
work seriously toward our goals. It wasn't easy – and we had
many obstacles to overcome.

To begin with, we had to upgrade Tavger's methods to mass-
production scale. This brought us face-to-face with our first and
most serious problem: obtaining the large number of required
snails. In Israel, the Society for the Protection of Nature had designated all sea invertebrates a protected species. In order to
obtain permission to collect small amounts of murex, we sought
the help of then-president Chaim Herzog, the son of Rabbi Isaac

Halevi Herzog the porphyrologist. That minimal number sufficed for experimental purposes, but for larger batches and actual manufacture, we needed another source.

Ari, a dentist, told one of his patients about our exciting venture as he worked on his teeth. The patient, once he could talk again, asked to see a picture of the snail and declared that he had seen those exact snails in fish markets in Greece. That was how on a business trip to Europe I came to stop in Pyrgos – asking for help from the old woman dressed in black, sewing a button on a shirt – and bought seven kilos of snails, the malodorous glands of which I smuggled back to Israel in a jar.

The Greek source lasted only a short while, though. The fishermen we hired proved unreliable, so we had to look for other options. In due course we found a supplier in the town of Jerez de la Frontera, near the ancient Phoenician city of Cádiz on the Atlantic coast of southern Spain. It was this place from which, at the turn of the seventeenth century, the English took the word *sherry,* the city's famous white wine, called in Spanish *vino de Xeres.* Jerez was one of the first cities in Spain to expel its Jews, even before the Catholic monarchs King Ferdinand II and Queen Isabella issued the Alhambra Decree in 1492.

There, putting history to rights in our small way, we arranged for a local group of fishermen to collect the snails, break open the shells, and send us just the glands. That channel proved excellent – until an oil spill west of Gibraltar resulted in a ban on the export of all seafood from the area. Eventually, we found a stable and lasting supplier along the shores of the Adriatic in Croatia.[164]

Over the years, the project of restoring the practice of dyeing *tekhelet* from the murex *hillazon* has grown far beyond what we could have imagined as we emerged from the water that winter day holding a big bag of snails. We moved the dyeing from our houses first to a tiny shack in the back of a local school and then to a small factory in the Judean desert on the road to the Dead Sea. Joined by talented and devoted companions, including Asaf Stein and Mois Navon, we now produce and distribute thousands of strings monthly. Awareness of the topic has increased tremendously as scholarly papers, research articles, educational materials, and books touching on all aspects of the subject continue to appear at a rapid pace.

It is no mere coincidence that the *tekhelet* string hangs from the hem at the corners of a garment. The blue thread emphasizes the importance of man's acceptance of his mission in life and of his commitment to moral and spiritual directives. The *tekhelet* thread literally and figuratively points the way to an ethical path, and a compass is most essential when a person is lost, unsure of what road to take or which direction to follow.

Such a situation often occurs on the border between events in life. A border signifies change, transition; it marks the end of one state, one status, and the beginning of another. The hem at the edge of a garment represents the boundary between what is inside, familiar, and comfortable and what is beyond, new, different, and uncertain. When people live in stability and calm, following their regular routines, they rarely encounter opportunities, ideas, or challenges that require a reevaluation of their fundamental principles. They don't have to choose between conflicting paths or to weigh the benefits or risks of divergent courses while lacking the necessary data and experience to decide wisely. But in that comfort zone, in the secure middle, lies stagnation; there is no room for change, improvement, or growth. To live a full life, a person must be ready to flee the interior and travel to the border – and even farther.

At the frontier, however, vulnerability exists. The unknown that lies beyond may not yield. Danger lurks there, and hidden obstacles may block the path to a remote, unachieved goal. The possibility of failure and disillusionment always accompanies the attempt to grow. Life is risk. But the string of *tekhelet* symbolizes the special guidance for those willing to confront doubt and uncertainty in order to flourish.[165]

For more than a millennium, no eye had seen threads of genuine *tekhelet*. The determination of a few individuals ready to embark on an uncertain quest, to confront risk and ridicule, to persist in the face of near-certain failure, allowed the ancient traditions to once again be realized.

Today hundreds of thousands around the world wear the *tekhelet* strings on their prayer shawls. To paraphrase the words of the ancient Midrash: Now we no longer have only white strings, for *tekhelet* once again has been revealed.

Epilogue
SOMETHING OLD, SOMETHING NEW

*E*arly in the morning, I am working at my computer in a hotel room in Croatia. I log in to Facebook and check the Phoenician Dyers page to see if anyone has posted anything new and exciting – perhaps an article relating to a recent archaeological find or maybe something about the discovery of a new medicinal use for a chemical found in a rare species of murex. This morning "Carthaginian Wanderer" has posted photos of his recent experiments in natural shellfish dyeing. The shots show tufts of wool, deep and lustrous blue, which he says were dyed using the eco-friendly method of milking the snails.

My wife and I have come to Croatia to meet the fisherman who provides the snails for our dyeing enterprise. In order to supply the growing demand for *tekhelet* strings, we need to drastically increase our stock and secure a steady supply of snails. Members of our group have traveled the Mediterranean looking for reliable murex dealers. Ari Greenspan and Joel Guberman have just returned from a trip to Djerba, a small island off the coast of southeast Tunisia, about halfway between the ancient Phoenician cities of Carthage and Sabratha.

It is here on the Dalmatian coast of Croatia that Cadmus, son of Phoenician king Agenor and brother of Phoenix, is said to have built a palace after being exiled from Thebes. Cadmus gets the mythological credit not only for founding that Greek city, but also for introducing the Phoenician alphabet to the Greeks. Sometimes it seems that we are following the same routes that those enterprising sea traders traveled three thousand years ago. We cannot escape them.

My life has certainly taken some unexpected turns. I make my living working in the high-tech communications industry — about as far removed from ancient dyeing techniques and contemporary religious innovations as you can get. In a spirit of lighthearted adventure, I decided, together with two childhood friends from New Jersey, to go scuba diving one day as a favor to some eccentric rabbi with a decidedly offbeat agenda. The journey that started that day led not only to ports around the Mediterranean but also to an engrossing, multifaceted intellectual adventure, to a previously unimagined wealth of knowledge and experience.

The world of the ancient murex dyers remains very much alive. Archaeologists are excavating the far-flung Phoenician trading posts, and in journals of ancient history around the world scholars are debating answers to questions about the structure of Phoenician ships. Our simple little sea snail becomes an amazingly complex creature when examined by expert marine biologists and by chemists who analyze the processes that result in dye precursors. Physicists and their arcane theories of color help to explain the scarcity of natural blue and the structure of the dye molecule. Ancient linguistics, color theory, mythology, as well as Jewish history, law, and customs round out the many varied disciplines that have engaged me. Each new fact or theory leads to another area that requires further exploration, and it has been exhilarating to constantly fit new pieces into a larger framework.

The knowledge that my research has accumulated represents only one facet of my personal growth. Another, in some ways more important, has come through making the acquaintance – both actual and virtual – of a variety of people connected in one way or another with shellfish dyeing. A worldwide camaraderie of individuals interested in aspects of the topic, ready to share their knowledge and enthusiasm, has gradually developed. I have corresponded and met with artists, chemists, professional and amateur dyers, Nobel Prize winners, politicians, professors specializing in fields of study from quantum chemistry to Hittitology, rabbis, snail milkers, Supreme Court justices, teachers, and other wonderful, interesting, and interested people from every corner of the globe who write with questions and comments on a daily basis.

Blue is no longer a scarce and valuable commodity; it is everywhere today and attracts little attention. But the ancient blue dyes can still stir strong feelings not only among Jews who have found authentic *tekhelet* again, but also among admirers of the skilled craftsmen of old whose achievements had such far-reaching effects.

My wife and I drive to the picturesque fishing town of Trogir, just north of Split, where the emperor Diocletian built his retirement palace. There we meet our friend Renato, the fisherman who supervises the collection of the snails. We set out on his boat to survey the rocky coast and the many small islands that serve as the habitat of the *Murex trunculus.*

We pull up next to a small skiff, and Renato introduces us to one of his local fishermen. The man reels in a rope that pulls up a wicker basket, which, he tells us, had been baited with chunks of raw fish and lowered to the seabed a few days earlier. As the basket emerges from the water, we see that it is covered with dozens of *trunculus* snails.

Ancient method of baited baskets used to catch snails, still used today
COURTESY OF E. SPANIER, KETER PUBLISHING HOUSE

The colorful account written some two thousand years ago by Pliny the Elder comes to mind. He describes how the snails are caught with baskets "cast into the sea, and in them cockles are put as bait…. In this way, victims to their own greediness, they [the snails] are drawn to the surface hanging by their tongue."[166] The experience stands as yet another example of what has become an integral part of my life's goal: the modern renewal of ancient techniques and a return to practices forgotten for so many centuries and nearly lost to history.

SOMETHING OLD, SOMETHING NEW

Much has changed over the years, and much stays the same. Gazing at the sun setting on the Adriatic, my mind lingers over the vast sky and the eternal sea, and my thoughts are inexorably drawn to threads of blue.

Acknowledgments

A book of this scope, touching as it does on so many time periods, disciplines, and areas of research, will inevitably reach beyond the expertise of any person. We are indebted to the real experts who reviewed particular chapters for accuracy and provided invaluable input. In particular, we thank: Dr. Donald T. Ariel, head, Coin Department, Israel Antiquities Authority; Dr. Kirsten Benkendorff, Marine Ecology Research Centre, Southern Cross University; Dr. Shulamit Eliash, General History Department, Bar-Ilan University; Dr. Wayne Horowitz, Department of Assyriology at the Hebrew University, Jerusalem; Professor Zvi Koren, Department of Chemical Engineering, Shenkar College of Engineering and Design; Professor Alan Kropf, Julian H. Gibbs professor of chemistry at Amherst College; Dr. Ilan Sharon, the Institute of Archaeology, the Hebrew University, Jerusalem; and Dr. Ari Zivotofsky, the Multidisciplinary Brain Research Center, Bar-Ilan University.

Special thanks to President Bougie Herzog and Kurt Raveh for the time they shared and the personal information they provided.

We also thank Rabbi Shlomoh Taitelbaum, Danny Oberman, Rabbi Michael Taubes, and Dr. David Matar, who read carefully through the manuscript, giving us important feedback and observations, and to Lev Roiterstein for his help in translating some of the Russian material.

Laurie Abkemeier, our agent, took us by the hand and guided us through the brave new world of book publishing. Without her direction and steady, cheerful encouragement, this book would never have gotten off the ground. We are indebted to her for her enthusiasm and invaluable input.

We thank the entire team at Globe Pequot for their careful proofreading and meticulous editing. We are especially grateful to our editor, the erudite James Jayo, who taught us what expert editing is all about. His detailed comments, suggestions, and substantial contributions, and his commitment to both the content and form of our work, significantly raised the professional level of this book.

A special thanks goes to our partners and friends – Dr. Ari Greenspan, Joel Guberman, and Rabbi Eliyahu Tavger. Our involvement with *tekhelet* has been one of the most important, engrossing, and fulfilling parts of our lives. The journey we have taken together has been exciting and fascinating, and we are grateful to have been able to share this adventure with them.

Lastly, we have no words to express our gratitude to Professor Leo Taubes, whose touch and influence appears on virtually every page. His broad knowledge, penetrating logic, sensitivity to style, and skill at organization proved indispensable in this book's development. What a joy and privilege to be able to work so closely on a project such as this with your father-in-law and father.

Endnotes

I: THAT DYE OF DYES

[1] This has been challenged, and some would give that credit to the Italian chemist, Bartolomeo Bizio. See F. Ghiretti, "Bartolomeo Bizio and the Rediscovery of Tyrian Purple."

[2] Ball, *Bright Earth,* 198.

[3] Throughout this book, we will describe the color of *tekhelet* as sky blue or azure. As will become clear, this is not a universally accepted position, and it certainly wasn't the consensus opinion among scholars before the 1980s. Some scholars maintain that the hue tended more toward violet, a blue purple. The depth of the color is also debated, with some claiming it was a darker shade of blue. The reasoning that leads to the opinion that *tekhelet* is the color of the clear, cloudless, midday sky will appear later in the book.

[4] The usual figure is something like 250,000 snails for an ounce of pure dye, based on the early twentieth-century research of chemists such as Paul Friedländer. Our numbers come from extensive experience with dyeing wool and probably fall closer to the numbers that ancient dyers would have realized.

5 Born, W. "Purple in Classical Antiquity," 114.

6 Esther 8:15. All subsequent translations of the Bible come from *The New JPS Translation According to the Traditional Hebrew Text*, The Jewish Publication Society, Philadelphia, 1985.

7 *Midrash Tanhuma,* Shelach.

II: OUT OF THE BLUE

8 The story of printing by movable type in some ways echoes that of shellfish dyeing – but in other ways opposes it. In the case of printing, the Minoan disc notwithstanding, that technology played no role in ancient society, while today it's hard to think of one more central to nearly every aspect of the modern world. Shellfish dyeing takes the opposite direction. It would be impossible to argue for the centrality of that art in today's society, yet in ancient times it acted as a cornerstone in the economic, aesthetic, and political arenas.

9 Purple even crept into legends about the exploits of King Minos. According to one Greek myth, he once besieged the city of Megara, ruled by a king named Nisus who had a magical source of strength that made him invincible. One lock of his hair was colored purple. As long as this hair remained on his head, he and his kingdom could not be destroyed. But much like the biblical Samson, deceived by his lover Delilah, Nisus, too, was felled by a woman. In this story, it was his own daughter Scylla, who, either for love (of Minos) or for money, depending on the version of the legend, cut off her father's purple lock and brought about the fall of his kingdom and the ruin of her city.

10 The Bible recounts that Cushan-Rishataim, ruler of Aram-Naharaim, sometimes identified as a Kassite king, conquered Israel

just after the death of Joshua, Moses's successor, and oppressed the Israelites for eight years (Judges 3:8).

[11] Genesis 10:9.

[12] I thank Professor Robert Stieglitz for directing me to his *Biblical Archaeologist* article in which he notes that small numbers of murex appear at the Early Minoan site of Myrtos (third millennium BCE), probably connected with diet rather than dye, although it is possible that Minoans made small quantities of the dye in the prepalatial period.

[13] The linguistic evolution of the Mesopotamian terms for "blue wool" makes for a fascinating study in itself. Professor Wayne Horowitz of the Hebrew University, an expert in Mesopotamian history and language, explained it to me as follows: "In Ancient Mesopotamia, there was no word for the color blue either in Sumerian or Akkadian. Hence, the Sumerian za.gin = *uqnû,* the word for lapis lazuli, was adopted to mean lapis lazuli–colored, i.e. blue and its various shades and hues. This was apparently first applied to the sky, and when blue wool reached Mesopotamia also to this product: sig.za.gin.na = *uqnâtu,* for blue-colored wool. In what is most likely a secondary development, at least for Mesopotamia, the foreign words *takiltu* for blue and dark blue, and *argamman,* were introduced. These two terms were rendered into Sumerian as za.gin.gi$_6$ = dark blue and za.gin.sa$_5$ = red-blue, i.e. purple. The earliest actual mention of blue wool (sig.za.gin) in the Land of Israel is in 'The Governor's Letter' from Aphek (around 1200 BCE)." See Horowitz, *Cuneiform in Canaan.*

[14] Numbers 15:37–40.

[15] Xun Zi, 8:23.

[16] Milgrom, *The Tassel and the Tallith*, 2.

[17] 1 Samuel 24:21.

[18] Exodus 19:6.

[19] Milgrom, *The Tassel and the Tallith*, 9.

[20] Ezekiel 23:6.

[21] Judges 5:30.

[22] Deuteronomy 33:19.

[23] See Jeremiah 10:8–9.

[24] Blum, *Purpur Als Statussymbol in Der Griechischen Welt*, 86.

[25] Itamar Singer, "Purple-Dyers in Lazpa," 22.

[26] Astour, "The Origin of the Terms Canaan, Phoenician and Purple."

[27] Proverbs 31:24.

[28] Zechariah 14:21.

[29] For this see Elayi, *The Phoenician Cities in the Persian Period,* 14, and also Stern, *Dor, Ruler of the Seas,* 21.

III: PURPLE PEOPLE

[30] Ezekiel, in his prophetic warning of impending destruction, captures the reputation of the Phoenicians as wealthy traders throughout the ancient Near East, dealing with a remarkable variety of goods and customers:

Tarshish traded with you because of your wealth of all kinds of goods; they bartered silver, iron, tin, and lead for your wares. Javan, Tubal, and Meshech – they were your merchants; they trafficked with you in human beings and copper utensils. From Beth-togarmah they bartered horses, horsemen, and mules for your wares. The people of Dedan were your merchants; many coastlands traded under your rule and rendered you tribute in ivory tusks and ebony. Aram traded with you because of your wealth of merchandise, dealing with you in turquoise, purple stuff, embroidery, fine linen, coral, and agate (27:12–16).

The long list continues: wheat, honey, oil, balm, wine, white wool, polished iron, cassia, calamus, saddlecloths, lambs, rams, goats, spices, precious stones, gold, choice fabrics, embroidered cloaks of blue, and many-colored carpets.

[31] This story appears on a wonderful coin minted in Tyre between the reigns of Elagabalus (218–222 CE) and Gallienus (253–268 CE). On the bottom of the coin a dog and a murex are portrayed, recounting the story of Melkarth and the discovery of the purple shellfish dye. On the top of the coin are the two Ambrosial rocks that flank an olive tree. The legend told by Nonnus from Panopolis in his *Dionysiaca* has these two rocks floating on the sea, and on one of the rocks an eagle perched upon a burning olive tree and a snake curled around it. Melkarth ordered his people to build boats to follow the rocks, and Tyre was founded where they came to rest. See Bijovsky, "The Ambrosial Rocks and the Sacred Precinct of Melqart in Tyre." I thank Donald Tzvi Ariel, head of the Coin Department for the Israel Antiquities Authority, for bringing this article to my attention.

Another depiction of this scene appears in *The Discovery of the Secret of Purple* by Peter Paul Rubens. The snail in that work is clearly a figment of the painter's imagination – certainly not a murex.

[32] Diodorus Siculus, *Bibliotheca historica,* bk. 20, lines 6–7.

[33] "Now I am sending you a skillful and intelligent man, my master Huram, the son of a Danite woman, his father a Tyrian. He is skilled at working in gold, silver, bronze, iron, precious stones, and wood; in purple, blue, and crimson yarn and in fine linen" (2 Chronicles 2:12–13).

[34] Ezekiel 27:7.

[35] Ezekiel 28:17.

[36] 2 Chronicles 36:18.

[37] Rogers, *Cuneiform Parallels to the Old Testament,* 316.

[38] In Plutarch's *Parallel Lives,* "Solon" bk. 5, sec. 2.

[39] Pliny the Elder, *The Natural History,* bk. 7, chap. 57.

[40] Homer, *Iliad,* bk. 13, lines 1–7.

[41] Jeremiah 4:13.

[42] Herodotus, *Histories,* bk. 4, lines 73–75.

[43] Polosmak, *Textiles from the "Frozen" Tombs in Gorny Altai 400-500 BC – An Integrate Study,* 222-223.

[44] Babylonian Talmud, Shabbat 26a, based on Jeremiah 52:16.

IV: EXPLORING DORA

[45] De Bourrienne and Phipps, *Memoirs of Napoleon Bonaparte,* vol. 1, chap. 19.

[46] Geologists debate the received wisdom that kurkar marks ancient coastlines. Optically stimulated luminescence, a method that measures the last time a substance was exposed to sunlight, provides information that indicates great variability in the formation dates of each ridge and no great differences between ridges.

[47] The Greeks extended their influence over Dor, though some eighteenth- and nineteenth-century scholars held that the reverse also happened. They claim that the founders of Paestum in Italy came from Dor, and the temples that were built there may have given the Doric name to one of the three orders of Greek architecture. Noted professor of architecture Richard Brown quotes this view in his *Domestic Architecture*, 62. The suggestion is older than that, as Brown mentions in a footnote, originally proposed by A. Mazzocchi in *Commentarii in regii Herculanensis Musei aeneas tabulas Heracleenses,* vols. 1 and 2 (Naples, 1754–55). We must take Mazzocchi's theory, however, with a grain of salt. Though one of the premier classicists and orientalists of the eighteenth century (one contemporary called him "the most learned Grecian of our time"), Mazzocchi was also a very proud Italian who sought to prove that the greatness of the Greeks came from their contacts with Italy, and the ancient Italians, in turn, directly descended from biblical migrations. There is no conclusive connection between Tel Dor and the Dorian Greeks or the Doric architectural order. S. R. Martin recently proposed a similar, though opposite, theory. The Hellenic inhabitants of Dor cleverly punned on the similarity of the Semitic "Dor" to the Greek Dorians, to manufacture a Greek ancestry.

[48] Conversation with the author.

[49] Ilan Sharon also questions this hypothesis. In correspondence he wrote: "What bothers me is the lack of evidence for heating/fermenting installations in that particular courtyard. I cannot escape the conclusion that this installation was used only for the first stage of the process (breaking the shells and treating the raw mollusk with quicklime) – and that further processing was done somewhere else."

[50] De Bourrienne and Phipps, *Memoirs of Napoleon Bonaparte,* vol. 1, chap. 19.

[51] Cvikel et al., "Napoleon Bonaparte's Adventure in Tantura Lagoon," 199.

[52] De Bourrienne and Phipps, *Memoirs of Napoleon Bonaparte,* vol. 1, chap. 19.

V: TRUE BLUE

[53] Babylonian Talmud, Baba Metziah 61b.

[54] Deuteronomy 22:11.

[55] *Midrash Sifre,* Shelach.

[56] Babylonian Talmud, Menachot 42b.

[57] Ibid.

[58] Ibid.

[59] The closest we have to eye-witness testimony are the words of the poet Yossi ben Yossi, who lived in Israel around the 5th century CE, when *tekhelet* was still extant. In his *Azkir Gevurot Eloha,* an acrostic portraying the Yom Kippur service, he describes the

High Priest: *"Atuy meil tekhelet, kezohar harakia* – Wrapped in a gown of *tekhelet* as lustrous as the sky." That characterization, comparing *tekhelet* to the sky, by one who presumably saw the actual colored wool, may be the most direct indication of the dye's hue. (Mirsky, *Piyutei Yossi ben Yossi*, 157)

[60] See for example – Rashi (1040-1105) on Numbers 15:38 who claims the color is green. A few sentences later (ibid 15:41), Rashi quotes Rav Moshe Hadarshan that the color of *tekhelet* is similar to the sky as it darkens towards evening, which would be a midnight blue or purple.

[61] There are some who disagree and actually arrive at the exact opposite conclusion. They maintain that *tekhelet* could not be the exact molecular equivalent of *kala ilan*, since the Talmud does propose that tests could distinguish between them. In my opinion, this argument is not credible. On the basis of current understanding of the dye chemistry and the standard interpretation of the Talmudic tests, those procedures would not cause any change in cloth dyed with plant based indigo, *kala ilan*, which was meant to fade under their influence. If one wants to uphold the chemical efficacy of the tests, one would have to posit that there are some yet-to-be-understood reactions that involve, perhaps, the methods by which the ancients dyed. If this is true, then one can take that reasoning further and apply it to murex dyeing as well. After all, there are significant amounts of many other substances along with indigo in murex-derived dye. Even if those additions are too small to be detected by the eye and change the perceived hue, they may play some little understood chemical role that is picked up via the Talmudic tests.

VI: THE MISSING SHADE OF BLUE

[62] Yadin, *The Finds from the Bar-Kokhba Period in the Caves of the Letters,* 182–87.

[63] Threads can be twisted then plied in different directions, as can be seen in the diagram. z-spun threads are considered "a manner of spinning that is very rare in Israel and Egypt. An examination of thousands of textiles throughout this region that was carried out by Shamir revealed that most textiles from the Roman period are s-spun and the few that are z-spun in both warp and weft are considered to be imported." (Sukenik, "Purple-Dyed Textiles from Wadi Murabba'at," 51, quoting Shamir 2006, 210-212)

In fact, the case of Textile 22 is more complicated since the threads are plyed, which is quite unique, and therefore comparison to other fabrics is inconclusive.

[64] Arrian of Nicomedia, *Anabasis,* bk. 6, chap. 29.

[65] Pliny the Elder, *Natural History,* bk. 9, chap. 36.

[66] In the realm of legend and folklore, there have been suggestions that the belief in vampires derives from cases of porphyria. In 1985, in a talk before the American Association for the Advancement of Science, biochemist David Dolphin offered a spine chilling suggestion. He proposed that "blood-drinking vampires

were in fact victims of porphyria trying to alleviate the symptoms of their dreadful disease." Afflicted individuals supplied their bodies with the heme molecules they lacked by sucking and ingesting human blood. The biochemist found support in the fact that porphyria can lead to photosensitivity, where exposure of the skin to light can cause great pain, which explains the vampire's aversion to sun. Porphyria can also lead to a tightening of the gums, tending to make the teeth protrude, like fangs. And what is known to exacerbate the symptoms? Garlic, of course!

[67] Reinhold, *History of Purple as a Status Symbol in Antiquity,* 71.

[68] Dio Cassius, *Roman History,* bk. 49.

[69] Seutonius, *De Vita Caesarum,* Nero, 32.

[70] Herzog and Spanier, *The Royal Purple,* 110.

[71] It is, perhaps, possible to find traces within the Talmud itself of the differing attitudes that the Roman authorities took towards enforcing the ban on *tekhelet* over various periods. An argument is recorded (Menachot 38a-39a) between Rabbi Yossi Haglili (1st-2nd century) who states, "if one has no *tekhelet*, he should put on white [threads on his tzitzit]," and Rabbi Yehuda Hanasi (d. 220) who is of the opinion that *tekhelet* and white are mutually dependent and the absence of one on the tzitzit invalidates the other. Baruch Chizik (*Otzar HaTzemachim*, Vol. 22) suggests that these positions represent two periods. Rabbi Yossi Haglili lived during the Kitos War (115-117) and the Roman general Lucius Quietus would have been vigilant in suppressing all forms of Jewish national expression including the wearing of *tekhelet*. Thus Rabbi Yossi takes a lenient position recognizing the dangers and hardship involved in procuring and displaying the *tekhelet* strings. By

the third century, however, with a thawing of tensions between Rome and Israel, and subsequent lack of enforcement of the restrictions on wearing *tekhelet*, Rabbi Yehuda Hanasi insists that people make the effort to obtain the precious strings and fulfil the commandment in complete compliance with the instructions of the Bible.

[72] Babylonian Talmud, Sanhedrin 12a.

[73] Eusebius Pamphilius, *Church History, Life of Constantine, Oration in Praise of Constantine*, Ch. XXXIII:3.

[74] Pharr, *Theodosian Code and Novels,* 288.

[75] One wonders whether Justinian's fascination with purple was influenced by his wife, Theodora. Early in his reign, Justinian found himself in an extremely precarious situation when attending a chariot race at the hippodrome of Constantinople. Enmity between two rival political factions (the Greens and – ironically – the Blues) erupted in the Nika riots. Justinian, fearing for his life, decided to flee. His stint as Emperor might have ended then and there had it not been for Theodora's inspiring words, an eloquent oration that changed the course of history:

My lords, the present occasion is too serious to allow me to follow the convention that a woman should not speak in a man's council. Those whose interests are threatened by extreme danger should think only of the wisest course of action, not of conventions. In my opinion, flight is not the right course, even if it should bring us to safety. It is impossible for a person, having been born into this world, not to die; but for one who has reigned it is intolerable to be a fugitive. May I never be deprived of this purple robe, and may I never see the day when those who meet me do not call me empress. If you wish to save yourself, my lord, there is no difficulty. We are rich; over there is the sea, and yonder are the ships. Yet reflect for a moment whether, when you have once escaped to a place of security, you would not gladly exchange such safety for death. As for me, I agree with the adage that the royal purple is the noblest shroud. (Safire, William, ed, *Lend Me Your Ears: Great Speeches in History,* W.W. Norton & Co., New York, 1992, p. 37)

[76] Babylonian Talmud, Menachot 43b.

[77] Babylonian Talmud, Menachot 42b.

[78] *Midrash Tanhuma,* Shelach.

[79] In a much later source, a passage appears that some use to claim that Jewish blue and purple dyeing continued long after this, namely the twelfth-century travel log of the intrepid Benjamin of Tudela. There is much confusion regarding what exactly Benjamin found when he visited Tyre. For example, the article on "Dye and Dyeing" in the *Jewish Encyclopedia,* first published in 1906, states: "In the twelfth century the Jews of Tyre were still purple dyers and manufacturers of glass (see Benjamin of Tudela, '*Travels,*' ed. Asher, p. 30b)." Rabbi Herzog also makes this assertion in *The Royal Purple* (p. 112). Asher indeed translates the log of Benjamin (*The Itinerary of Rabbi Benjamin of Tudela,* vol. 1 [New York: Hakesheth, 1840], 63) as follows: "The Jews of Tsour [Tyre] are shipowners and manufacturers of the far-renowned tyrian glass, the purple dye is also found in this vicinity."

But Adler's translation (1907) has "The Jews own sea-going vessels, and there are glass makers amongst them who make that fine Tyrian glass-ware which is prized in all countries. In the vicinity is found sugar of a high-class, for men plant it here, and people come from all lands to buy it." We have sugar now instead of purple dye. The Hebrew original isn't clear either, with various manuscripts giving *asukar, hasikar,* or *hatzukru.* Asher's Hebrew manuscript reads "*hasikar.*" Both he and Rabbi Herzog translate that word as coming from *sikra,* meaning red, and render that as purple, perhaps influenced by the Tyrian connection. But looking at the other versions and manuscripts convinces me that Adler's translation is correct, and in the twelfth century the Jews of Tyre dealt in sugar, not in snails.

VII: MOOD INDIGO

[80] Dendel, *You Cannot Unsneeze a Sneeze and Other Tales from Liberia.*

[81] The story also turns out to reveal a very real, though extremely rare, physiological phenomenon: Blue Diaper Syndrome. The disorder involves the metabolism of tryptophan, an amino acid essential for the human diet. According to an article in the American Journal of Medicine, "Bacterial degradation of the tryptophan leads to excessive indole production and thus to indicanuria which, on oxidation to indigo blue, causes a peculiar bluish discoloration of the diaper." Indicanuria is the excretion through urine of indican, a precursor of indigo dye that turns blue on exposure to oxygen in the air. [Drummond, "The Blue Diaper Syndrome"].

[82] Julius Caesar, *The Gallic Wars*, bk. 5, chap. 14.

[83] E-mail correspondence with the author.

[84] Balfour-Paul, *Indigo*, 127.

[85] Ibid., 100–102.

86 Sandberg, *Indigo Textiles,* 19. Other descriptions of this process are less cheerful. Balfour-Paul, for example, writes: "Groups of near-naked men and women had to walk up and down waist deep in the slimy liquid for several hours, beating it with implements such as wooden paddles, or even with their bare hands…. [The tanks were labeled] 'devil's tank', as the terrible fumes… 'killed many workers.' Local populations were understandably reluctant to undertake a job said to cause, if not death itself, at least cancer, impotence, headaches, and temporary lameness" (Balfour-Paul, *Indigo,* 110–11).

87 Bemiss, *The Dyer's Companion.*

88 Edmonds, *The History of Woad and the Medieval Woad Vat,* 26.

VIII: THE QUEST FOR THE HOLY SNAIL

89 In the narrative that follows I have drawn on the works on the history and teachings of Ishbitz/Radzyn by Shlomo Zalman Shragai, an Israeli politician, the first democratically elected mayor of Jerusalem (1950–52), and a devout Hasid of Radzyn.

90 Morris Faierstein in *All Is in the Hands of Heaven,* on Mordechai Yosef Leiner's teachings, includes an appendix entitled "The Friday Night Incident in Kotsk: History of a Legend." The chapter begins: "The antinomian legend of the 'Friday night incident' in Kotsk is one of the best-known Hasidic tales. Its truth or falsehood has been discussed for the last sixty years." Faierstein attempts to trace the historical development of the story and to assess whether it was accurate and even plausible.

In the version that I present, I have taken parts of the story from Martin Buber's account in *Der Grosse Maggid und Seine Nachfolge* as well as from the memoirs of a member of the Leiner family, *Dor Yesharim,* originally published in 1925 and reprinted in S. Z. Shragai's *Bmaayanei Chasidut Ishbitz-Radzyn.*

Some scholars and many Hasidim argue that the incident never took place. If it did happen, Leiner is said to have played a central part, though the exact nature of his role is debated. What is certainly true is that Leiner broke with Menachem Mendel around the time that the latter went into seclusion, and that a feud ensued between subsequent generations of Kotsker and Radzyn Hasidim.

[91] Leiner, *Ptil Tekhelet*, 8.

[92] Babylonian Talmud, Menachot 44a.

[93] *Shir HaShirim Rabbah*, 4:23.

[94] Leiner, *Sefunei Temunei Chol*, 33, based on the Mishna *Kelim*, 12:1.

[95] Maimonides, *Mishneh Torah*, Hilchot Tzitzit, 2:1.

[96] Babylonian Talmud, Shabbat 75a.

[97] Shragai, *Bi-netive Hasidut Izbitza-Radzin*, vol. 2, 194, 198. See also the poem "The Song of the Radzyner" written in 1943 by Yitzchok Katzenelson.

[98] See Gadi Sagiv, *A New Perspective on the Tekhelet Controversy of the Late Nineteenth Century*, Zion, Vol. 82(1), 2017 pp.59-95. (Hebrew)

IX: BLUE BLOOD

[99] One of the other rabbis from whom Herzog received ordination was the world-renowned Rabbi Jacob Vilkowsky, whom he saw as his ultimate mentor and for whom he named his second son, Jacob.

[100] Herzog's relationship with Sinn Féin rebels and the influence that his friendship with de Valera had on his intellectual outlook offers a fascinating study in itself. Both men dealt with the complex interplay of state and religion, and how to ensure the rights of the minority while allowing for a country with a specific national religious character. See Matveev, "The Rebbe of Sinn Féin."

[101] Rabbi Herzog's grandson and namesake, Isaac Herzog (nicknamed Bougie in Israel), the head Israel's Labor party and Leader of the Opposition, has earned a reputation for being an advocate for the rights of the disadvantaged and a defender of the working class. "Though my grandfather died a year before I was born," he told me over a cup of green tea in a Tel Aviv cafe, "his values and ideals permeated our house and made a deep impression on me. His attitudes towards religion and state were greatly influenced by his years in Dublin watching the nascent country fight for and gain its independence. He was a fervent Zionist who believed in harmonizing Judaism and democracy, and together with my grandmother Sarah built a home based on charity and kindness. These values were absorbed by my father, Chaim, who served as a general in the Israeli army, ambassador to the United Nations, and later as Israel's sixth president. His feelings of compassion towards others, in my opinion, were amplified during his

years as a British officer in World War II, where he saw the horrors of war firsthand. I grew up in a family of people who care for people."

102 Shaul Meislish, "Toldot Harav Herzog," in *Maśu'ah Le-Yitshak,* ed. Eliash, Warhaftig, and Dosberg, 19.

103 A number of versions of this story exist. Rabbi Herzog's son Yaakov, in an interview to the radio station Galei Zahal in November 1969, asserted that it was indeed President Roosevelt who asked his father not to return to Israel. Others maintained that it was Lord Halifax who sent a messenger to Rabbi Herzog explaining that, in light of Rommel's advance, the British government was considering evacuating their citizens from the Middle East. Halifax counseled that Rabbi Herzog, as a British citizen, should stay in America. Still another version places the conversation between Rabbi Herzog and the mayor of New York City, Fiorello La Guardia. For a full discussion see the article by Ari Shvat, "Rabbi Herzog's Certainty That There Won't Occur a Third Destruction," in *Maśu'ah Le-Yitshak,* ed. Eliash, Warhaftig, and Dosberg, *Masua Le'Yitzchak, Part II*

104 McCullough, *Truman,* 620.

105 Herzog and Spanier, *The Royal Purple,* 116.

106 Ibid.

107 Ibid., 131 n. 410.

108 Professor Mary Orna kindly pointed out that Prussian blue served as the key clue in solving another case of ancient authenticity. "Archaic Mark" as it became known, or MS 2427 in the scholarly jargon, is a beautifully illuminated miniature manuscript

of the Gospel of Mark, presumed to date from the thirteenth century. That's certainly what the University of Chicago thought when it acquired the codex in 1937. But since that time, scholars have argued about it. Some believe it to belong to Category I codices (that is, Alexandrian-type texts usually from the fourth century or earlier). Others had doubts. In 1988 Professor Orna tested the pigment contained in one of the illustrations and found it to include none other than Prussian blue, proving that the codex couldn't be any older than 1704, when Diesbach discovered it. Subsequent textual investigation by Stephen Carlson and more detailed chemical analysis in 2006 by Mitchell, Barabe, and Quandt confirmed it to be a nineteenth- or even early twentieth-century forgery.

[109] Bartoll, "The Early Use of Prussian Blue in Paintings."

[110] See Yaakov Leiner, "Mifal Tzviat HaTekhelet BMidinat Yisrael," in *HaTekhelet*, Burshtein, 59–63. See also the chapter "Hidush HaTekhelet B'Medinat Yisrael," 197–204.

[111] As Herzog put it, "Science knows nothing of such a septuagenarian 'appearance' of any of the denizens of the sea" (Herzog and Spanier, *The Royal Purple*, 69).

[112] Herzog, "HaTekhelet B'Yisrael" 11:2, in *HaTekhelet*, Burshtein, 422.

[113] Babylonian Talmud, Shabbat 26a.

[114] *Shir HaShirim Rabbah*, 4:23.

[115] I thank Rabbi Dr. Chaim Twerski for sharing this letter with me.

[116] Edelstein himself published another important manuscript – namely Rabbi Herzog's dissertation. Through Edelstein's funding, librarian Moshe Ron took on the project of transcribing, translating, and editing the handwritten work. It appeared in 1987 in book form, edited by Ehud Spanier, together with many articles describing the current state of *tekhelet* research.

[117] This is one version of the apocryphal stories of the serendipity that led to Elsner's discovery of the photochemical processes that debrominate snail dye. It may not have been completely accidental, though, and Elsner may have been looking for such effects. As an experienced indigo dyer, he knew that dyeing in direct sunlight could change the color somewhat. He also may have seen the paper by Driessen, who first suggested this phenomenon (Driessen, "Uber Eine Charakterische Reaktion Des Antiken Purpurs Auf Der Faser").

[118] Some chemists and archaeologists believe that ancient dyers wouldn't have produced sky blue and wouldn't have figured out how to obtain that color. They argue that, since the fermentation process, which is necessary to make the murex extract suitable for dyeing, is more efficient in oxygen-poor conditions, the dye vat had to be kept closed and away from direct sunlight. Dyers, therefore, wouldn't have observed the effects of photo-debromination turning purple to blue, accomplished as it is in the open air and bright sunlight.

I find this position highly implausible since we have discovered many ways to achieve azure blue from the murex dye, whether through sunlight or steam, and I am quite sure that ancient dyers practicing their craft over thousands of years would have noticed this phenomenon. Furthermore, the Pazyryk and

Murabba'at fabrics clearly demonstrate that ancient dyers could and did dye blue with murex. To be fair, though, modern researchers didn't know about the photo-debromination until Elsner's discovery.

X: THE SELFISH SHELLFISH

[119] Dr. Harry G. Lee kindly pointed out the passage from Linnaeus and its interpretation. In an e-mail to me, Dr. Lee wrote:

The species epithet was employed as an appositional noun. We know this because Linnaeus consistently initiated species epithets with upper case if nominative and lower case if adjectival. Trunculus is the diminutive of truncus, and both are used as nouns or adjectives in Latin. In the context of Linnaeus's concept of Murex, one meaning of trunculus: "little cut-off," seems apt. As I look through the taxa the master considered congeners, many of them have much longer siphonal canals. He wrote "M[urex] testa ovata nodosa anterius [sic] spinis, **cauda breviora truncata** *perforata" [my boldface] (Linnaeus 1758:747). I think he was trying to capture the chopped-off canal (cauda).*

Though Dr. Lee is probably correct in attributing the "truncation" that Linnaeus mentions to the canal (*cauda* means tail), I chose to ascribe that appellation to the spines.The correspondence to the colloquial Hebrew was too good to pass over. In addition, Linnaeus does speak of *spinis,* which could mean spine or prickle, so I felt license was justified.

[120] The murex also secrete choline esters, which are muscle relaxants. Injecting these into their prey helps the murex subdue their victims for long periods. The esters also help relax the adductor

muscle in bivalves for the murex to gain access to the meat without having to bore through the shell. Dr. Benkendorff observed that these chemicals make abalone loosen their grip on the substrate to which they had attached, so the snails can flip them over and feast en masse.

[121] Indole forms from a six-sided benzene ring attached to a five-membered pyrrole ring. Used since the fifteenth century, benzene was originally known as "frankincense of Java." The chemical formula for benzene, namely its six carbon atoms, was understood early on, but the actual structure remained a puzzle for many years. German chemist Friedrich Kekulé first proposed the ring formation in 1865 after waking from a dream in which he saw a snake eating its own tail.

[122] The murex family, through multiple precursor pathways, actually produces a number of dyes that fall into three general categories based on the fundamental molecules: indigo, indirubin, and isatin. The first two have the same molecular formula; that is, they have the same types and amounts of each atom, but they differ structurally. These types of molecules are called isomers. Where indigo is aligned linearly, indirubin is L shaped. The third, isatin, is essentially an oxygenated version of indole or half an indigo molecule. In each of these groups, additional dye molecules form when bromine replaces one or two of the hydrogen atoms. Since indirubin and isatin are not symmetrical, the position of the bromine atom makes a difference. The following table illustrates the ten dye molecules that the murex can produce.

Structure	Atomic Component		Common Name	Color
Indigoids	X_6	$X_{6'}$		
	H	H	Indigo	Blue
	Br	H	6-Monobromoindigo	violet
	Br	Br	6-6'-Dibromoindigo	Reddish-Purple
Indirubinoids	X_6	$X_{6'}$		
	H	H	Indirubin	Various
	Br	H	6-Monobromoindirubin	Shades
	H	Br	6'-Monobromoindirubin	of
	Br	Br	6-6'-Dibromoindirubin	Maroon
Isatinoids	X_4	X_6		
	H	H	Isatin	Shades
	Br	H	4-Bromoisatin	of
	H	Br	6-Bromoisatin	Yellow

This information comes from data in the articles by Koren cited in the bibliography and is based mostly on the table in his article from 2006, *HPLC-PDA analysis of brominated indirubinoid, indigoid, and isatinoid dyes.*

[123] Aristotle, *The History of Animals,* bk. 5, part 15 (translated by D'Arcy Wentworth Thompson).

[124] Babylonian Talmud, Shabbat 75a.

[125] Dr. Kirsten Benkendorff, of Southern Cross University, confirms that the synthesis of the dye precursors – including bromoperoxidase activity – does indeed increase before and during mating.

[126] Earlier research into the correlation of dye production with snail sex claimed the opposite. See, for example, Elsner, "Solution of the Enigmas of Dyeing Tyrian Purple and the Biblical Tekhelet." Benkendorff's experiments appear more conclusive,

though more research is necessary to complete our understanding.

127 Professor Ehud Spanier from Haifa University suggested a different use that the murex might make of tyriverdin, namely as a pheromone. He proposed that the molecule could signal a chemical call to other murex to join together for mating and egg laying. Murex do tend to gather and lay their eggs en masse. This may also explain why males and not only egg-laying females produce the substance. Again, more research is necessary to confirm this hypothesis.

128 Professor Skaltsounis at the University of Athens School of Pharmacy is carrying out this work.

129 Babylonian Talmud, Bava Metziah 61b.

XI: DYED IN THE WOOL

130 Jerusalem Talmud, Shabbat, chap. 1, Halacha 3. See the commentary of the Pnei Moshe.

131 Acts 16:11–15.

132 Faber, *Dyeing and Tanning in Classical Antiquity,* 285.

133 I thank Georg Stark, master blue dyer, for providing this information. By the way, here's a tip for gamers of the popular online RuneScape fantasy world: Aggie the witch can provide you with blue dye.

134 Pliny the Elder, *Natural History,* bk. 9, chap. 62.

135 Babylonian Talmud, Menachot 42b.

[136] Elsner, "The Past, Present and Future of Tekhelet," 175.

[137] Edmonds published two books: *Chaucer in Modern English Prose: The Canterbury Tales* and *Chaucer's Other Works: In Modern English Prose.*

XII: BEHIND BLUE EYES

[138] This is not strictly true. There are five ways to turn white light into blue. There are other ways to make blue in the universe. Some have to do with radiation – that is, generating light – which can be preferentially blue. A gas flame is blue. The water surrounding nuclear reactors gives off a beautiful blue in a process known as Cherenkov radiation. Kurt Nassau's *The Physics and Chemistry of Color* details all fifteen ways that color can be generated.

[139] When the revolution of modern physics came about, British poet J. C. Collins extended the poem with:

> *It did not last: the devil, shouting "Ho.*
> *Let Einstein be," restored the status quo.*

[140] The cones, however, aren't evenly distributed; about 64 percent are red or L cones, 32 percent are green or M, and only 2 percent are blue or S. Though these are so much scarcer, they are the most acutely sensitive to light. Humans can detect blue as effectively as red, but the lower density of the S cones means that the spatial resolution of blue is less than that of red.

Another, perhaps related, observation (for which I thank Professor Ari Zivotofsky) has to do with the plane of focus of the eye. Since, as Newton pointed out in his experiments on "re-frangibility," different colors bend to different degrees through prisms and lenses, the eye focuses red at a slightly different plane than blue. The retina is placed for optimal focus of L and M cones, namely red and green, so all the blue that we humans see blurs slightly. The blue that we do see, blurred or not, comes primarily from the indiscriminate sky and ocean, where sharp lines don't play a prominent role. It appears that, in the evolution of the human visual system, blue objects, so rare in nature, didn't present a significant requirement that needed accommodating.

[141] Humans who have one or two types of cone are said to be color-blind. Women have a higher sensitivity to color, for which there may be a physiological basis. As a recessive gene on the X chromosome, color blindness affects mostly men; 8 percent of the male population has some form of color blindness, whereas in women the frequency measures ten times less. In some rare cases, women develop a fourth cone (sensitive in the yellow region) that gives these lucky individuals even greater color perception. Gender-related genetic attributes may or may not have to do with women showing a greater faculty for color discrimination than men and girls learning to identify different colors earlier than boys.

There is an extremely rare type of color blindness, known as achromatopsia, where the person has no cones, only rods, and can't see any color at all. This condition is described in Oliver Sack's wonderful book, *The Island of the Colorblind*.

[142] The interaction of light and matter on the most fundamental level occurs when a photon strikes an atom or molecule. In order for the photon to be absorbed, thereby transferring its energy to

the atom or molecule, those particles have to be able to "do something" with the photon's energy. Molecules can absorb photons and start to rotate and vibrate, but the wavelengths that excite those modes are much longer than visible light, so all the energy transitions that take place in that range play no part in our perception of color.

According to quantum mechanics, molecules and atoms never stop moving – but they're not free to move at random; they must occupy specific and discrete energy states. In their ground state, they may rotate at specific speeds and specific energies, but when hit by the right photon, they can jump from a lower to a higher level. That happens only when the energy jump from one state to another precisely matches the energy of an incoming photon, as determined by the photon's wavelength. In that case, the molecule can absorb the photon and change its rotational speed. When it falls back to a lower energy level, the molecule slows down, moving from the higher energy state to a lower state in a single jump, emitting a photon with exactly the wavelength that corresponds to the drop in energy. This is the process that takes place, for example, in a microwave oven, in which the radiation corresponds to the wavelength that induces water molecules to rotate. The water in your food absorbs those photons and spins, bumping into the other food molecules along the way, thereby heating them up. The wavelengths that induce rotational states measure thousands of times longer than visible light (microwave ovens emitting a wavelength of about 12 cm), and so our eyes can't see them.

At shorter wavelengths and higher energies, the absorption of photons by molecules causes the atoms in the molecule to vibrate. Again, quantum mechanics constrains this movement to discrete energy states, and moving from one vibrational mode to another involves a jump in energy defined by the precise wavelength of the photon required to excite that mode. The wavelengths involved in vibrational transitions, though shorter than those needed to excite rotational states, are still longer than visible light and fall within the infrared region.

[143] To explain this a little more, let's use, as an analogy for a molecule, atomic balls held together by springs. Changing two factors can speed up the vibration: lighter balls or tighter springs. Bowling balls on a long loose spring will vibrate more slowly than Ping-Pong balls on a short tight spring. For vibrational energy to move up toward the visible, you need the lightest ball, hydrogen. The water molecule resembles Mickey Mouse, with an oxygen head and two hydrogen ears. A curious characteristic of the water molecule – discussed in Michael Brooks's 2008 book, *13 Things That Don't Make Sense* – is that even in its natural stable state it can form chains and structures, with tens or even hundreds of water molecules coming together. These clusters tend to "tighten the spring" and drive the vibrational energy a bit closer to visible – but still not close enough.

To get there requires harmonic overtones of the vibration, a property familiar to musicians. Pluck a taut string, and you hear a tone that comes mostly from its vibration at a fundamental frequency, with an added richness from higher harmonics (twice the fundamental frequency, thrice, etc.). Suppressing the fundamental frequency allows those higher harmonics themselves to be heard clearly; guitarists do this by lightly touching a string at the twelfth fret, which shortens the string by a half, as in the classic introduction to the song "Roundabout" by Yes.

Putting all these processes together, the vibrational energy of the higher harmonics of a water cluster just barely reaches the wavelengths of visible light. As a result, water slightly absorbs red light just enough to make the ocean look blue.

[144] Occasionally, however, along with absorption in the blue region, a molecule can have a rogue electron transition that also absorbs red. The chlorophyll in plants, for instance, absorbs both blue and red, and this absorption results in the lush green of grass and leaves.

[145] As reported in Wouters and Verhecken, "High-Performance Liquid Chromatography of Blue and Purple Indigoid Natural Dyes." Though it must be noted that other experiments give a slightly different absorption peak that seems to depend on the solvent used.

Six hundred thirteen also happens to be an important number in Jewish lore, representing precisely the number of commandments that tradition claims the Bible contains. Furthermore, letters of the Hebrew alphabet – like those of Arabic, ancient Greek, and Latin – were also used as numbers. A subset of what we call numerology, the practice of gematria, or isopsephy, as the Greeks called it, involves calculating the numerical value of words. Words with the same numerical value share an affinity with each other, thought to convey a hidden idea or some significant relationship that exists between the word and its number value. According to some scholars, the Hebrew word *tzitzit* – the fringes on the corners of a garment – taken together with the number of knots and strings, yields a value of 613. Wearing the tzitzit therefore numerologically reminds all who wear them to observe all 613 commandments.

That the *tekhelet* molecule, indigo, gets its deep sky-blue color from a strong absorption peak centered exactly at 613 nanometers makes for a remarkable coincidence linking ancient tradition and modern chemistry.

[146] Dye chemists assert that the part of the indigo molecule responsible for its color lies in the center and refer to it as the H-chromophore, due to its H-like shape. Proof that symmetry plays an important role in bringing the absorption spectrum of indigo to such uniquely long wavelengths comes from examining the property of its isomer, indirubin. That molecule has an identical atomic formula and consists of the same atoms, but its geometric structure is L shaped. The letter *H* has more axes of symmetry than the letter *L*, which must rotate a full 360 degrees to return to its original position. Changes in the electron configurations of indirubin require greater energy so that its absorption of light, more typical than that of indigo, falls in the higher energy blue end of the spectrum, giving it a more earthy, less heavenly red tone.

XIII: RHAPSODY IN BLUE

[147] Esther 1:6.

[148] Finlay, "Weaving the Rainbow," 401.

[149] Goethe, *Goethe's Theory Of Colours,* 329.

[150] Bush and Shih, *Early Chinese Texts on Painting*, quoted in Finlay, "Weaving the Rainbow," 410.

[151] The blue and white of the Israeli flag were inspired by the *tekhelet* and white tzitzit of the Jewish prayer shawl. The powerful symbolism that Rabbi Soloveitchik describes, where the *tekhelet* reminds one of the unpredictable course that history often takes, is especially poignant with respect to Jewish history, and the blue in the flag evokes the mystery of the events and processes that ultimately led to the reestablishment of a Jewish homeland after 2000 years of exile. For a thoughtful interpretation of the relationship between *tekhelet* and the Israeli flag, see the play, "The Flag that Came Out of the Blue," in *Old Wine, New Flasks*, by Roald Hoffmann and Shira Leibowitz Schmidt. A fascinating history of the Israeli flag can be found in Dani Baron's *Ish al Diglo*.

[152] Sagiv, "Dazzling Blue: Color Symbolism, Kabbalistic Myth, and the Evil Eye in Judaism" 59-95.

[153] Blue light can also treat skin conditions such as acne. Recent studies, however, advise extreme caution regarding overexposure to intense blue light, which seems to induce photooxidative damage that leads to age-related macular degeneration.

[154] Sacks, Oliver, Altered States; Self-experiments in chemistry, New Yorker, August 27, 2012

[155] Klein, *The Chelsea Hotel Manifesto*, 1961. Quoted from the Yves Klein Archives website.

[156] Kandinsky, *On the Spiritual in Art*, 64

[157] Cotel never lacked inspiration for his musical pieces. In his obituary, the *New York Times* recounted that

In 1996, while he was at his piano playing Bach's Well-Tempered Clavier, his 3-year-old cat, Ketzel, pounced on the keyboard. The professor grabbed a pencil and inscribed a descending paw pattern from treble to bass. A year later, he entered the score — if one can call it that — in the Paris New Music Review's One-Minute Competition, open to pieces of no more than 60 seconds. The judges gave Ketzel an honorable mention (www.nytimes.com/2008/11/03/arts/music/03cotel.html).

[158] From a personal correspondence.

[159] Galileo, *Il Saggitore,* trans. A. C. Danto in Columbia College, *Introduction to Contemporary Civilization in the West,* 789.

[160] Hyman, *The Objective Eye,* 14.

[161] Hoeppe, *Why the Sky Is Blue,* 13–14.

[162] Nabokov, *Laughter in the Dark,* 291. It is not, of course, Nabokov himself who speaks the words, but rather his character.

XIV: TANGLED UP IN BLUE

[163] The notion that artistic or aesthetic motivations enter into what many view as a cold, rational, legalistic realm of Jewish law may seem strange, but that appears to be the case with regards to *tekhelet.* For example, Abraham ben David, a leading twelfth-century legal scholar from Posquières in France, chose a specific method of tying that "is arranged in a very pleasing arrangement." Note also the instructions of Rav in the Talmud concerning the most attractive ratio for the tzitzit. See Navon, "Rav's Beautiful Ratio"

164 Two options for a sustained supply of murex dye remain to be investigated. The first, growing or farming the snails, our group has looked into on a very cursory level, having commissioned some research at the University of Miami into the conditions and parameters required. Another alternative is to develop techniques for eco-friendly extraction of the dye – that is, "milking" the snails – on an industrial scale. This would be the ideal method of choice for many reasons, and it is my personal hope that such a process comes to pass.

165 I heard this philosophical insight from the Radzyner Hasid who is currently responsible for dyeing the cuttlefish-based *tekhelet*. He attributed this idea to Gershon Henokh himself. How fitting and appropriate a thought for the Radzyner rebbe, who embodied the idea of risking everything in order to achieve one's mission, and who stands as a role model to all in this respect.

166 Pliny the Elder, *Natural History,* bk. 9, chap. 37

Bibliography

Abrahams, D. H., and S. M. Edelstein. "A New Method for the Analysis of Ancient Dyed Textiles." *American Dyestuff Reporter* 53 (1964): 19–25.

Adrosko, Rita J., and Margaret Smith Furry. *Natural Dyes and Home Dyeing.* New York: Dover Publications, 1971.

Adrosko, Rita J. "Natural Dyes in the United States." *United States National Museum Bulletin* 281 (1968): 17–19.

Allan, J. K. "Tyrian Purple: An Ancient Industry." *Australian Museum Magazine* 5, no. 5 (1934): 147–51.

Amar, Zohar. *Ha'Argamman: Purphura ve'Arjawan B'mikorot Yisrael ve'od Beirurim B'Inyanei HaTekhelet,* Mekhon Har berakhah, 5774, 2014

Anderson, D. T. "The Life Histories of Marine Prosobranch Gastropods." *J Malac Soc* 4 (1959).

Ariel, Shmuel. "Tekhelet B'Tzitzit: Mitzva Min Ha'Muvchar O Chiyuv Gamur?" *Techumin* 21 (2001): 475–85.

Aristotle, and D'Arcy Wentworth Thompson. *History of Animals.* [Belle Fourche]: NuVision Publications, 2004.

Astour, Michael C. "The Origin of the Terms Canaan, Phoenician and Purple." *Journal of Near Eastern Studies* 24 (1965): 346–50.

Aynard, J. M. "Coquillages Mésopotamiens." *Syria* 43 (1966): 21–37.

Baginski, A., and A. Tidhar. *Textiles from Egypt: 4th-13th Centuries*

CE. Jerusalem: L. A. Mayer Memorial Institute for Islamic Art, 1980.

Baker, J. T., and C. C. Duke. "Chemistry of the Indoleninones. II. Isolation from the Hypobranchial Glands of Marine Molluscs of 6-Bromo-2,2-Dimethylthioindolin-3-One and 6-Bromo-2-Methylthioindoleninone as Alternative Precursors to Tyrian Purple." *Australian Journal of Chemistry* 26 (1973): 2153–57.

——. "Isolation from the Hypobranchial Glands of Marine Molluscs of 6-Bromo-2,2-Dimethylthioindolin-3-One and 6-Bromo-2-Methylthioindoleninone as Alternative Precursors to Tyrian Purple." *Tetrahedron Letters* 27 (1973): 2481–82.

——. "Isolation of Choline and Choline Ester Salts of Tyrindoxyl Sulphate from the Marine Molluscs Dicathais Orbita and Mancinella Keineri." *Tetrahedron Letters* 15 (1976): 1233–34.

Baker, J. T., and M. D. Sutherland. "Pigments of Marine Animals VIII. Precursors of 6,6'-Dibromoindigotin (Tyrian Purple) from the Mollusc Dicathais Orbita Gmelin." *Tetrahedron Letters* 1 (1968): 43–46.

——. "Precursors of Tyrian Purple." *Food-Drugs Sea,* 1976, 345–53.

——. "Some Metabolites from Australian Marine Organisms." *Pure & Appl. Chem.* 48 (1976): 35–44.

——. "Tyrian Purple: An Ancient Dye, a Modern Problem." *Endeavour* 33, no. 118 (1974): 11–17.

Balfour-Paul, Jenny. *Indigo.* Chicago: Fitzroy Dearborn, 2000.

——. "Indigo – An Arab Curiosity and Its Omani Variations." In *Oman: Economic, Social and Strategic Developments,* edited by B. R. Pridham, 79–93. London: Croom Helm, 1987.

——. "Indigo in the Arab World." *Dyes in History and Archaeology* 9 (1990): 3–6.

—. "The Indigo Industry of the Yemen." In *Arabian Studies,* edited by R. B. Serjeant and R. L. Bidwell, 39–62. Cambridge: Cambridge University Press, 1990.

Ball, Philip. *Bright Earth: Art and the Invention of Color.* New York: Farrar, Straus and Giroux, 2002.

Baron, Dani. *Ish al Diglo: Masa Ishi B'Ikvot HaMoreshet HaHistorit shel Degel Yisrael,* Jerusalem, Steimatzky, 2016.

Bartoll, Jens. "The Early Use of Prussian Blue in Paintings." In *Art2008 9th International Conference, JerusalemIsrael,* proceedings of international conference on "Non-destructive Investigations and Microanalysis for the Diagnostics and Conservation of the Cultural and Environmental Heritage," 2008.

Bednarz, Terri. "Lydia Speaks: Examining the Life of Lydia through Her Social and Theological Context." Master's thesis, the Catholic Theological Union at Chicago, 2002.

Belsky, Judy. *Thread of Blue: A Journey through Loss, Faith and Renewal.* Southfield, MI: Targum, 2003.

Bemiss, Elijah. *The Dyer's Companion.* 2nd ed. New York: Evert Duyckinck, 1815.

Benkendorff, Kirsten. "Bioactive Molluscan Resources and Their Conservation: Biological and Chemical Studies on the Egg Masses of Marine Molluscs." PhD diss., University of Wollongong, 1999.

—. "Molluscan Biological and Chemical Diversity: Secondary Metabolites and Medicinal Resources Produced by Marine Molluscs." *Biological Reviews,* 2010, 1–21. doi:10.1111/j.1469-185X.2010.00124.x.

Benkendorff, Kirsten, Cassandra M. McIver, and Catherine A. Abbott. "Bioactivity of the Murex Homeopathic Remedy

BIBLIOGRAPHY

and of Extracts from an Australian Muricid Mollusc against Human Cancer Cells." *Evidence-Based Complementary and Alternative Medicine* 2011 (2011): 1–12.

Benkendorff, Kirsten, John B. Bremner, and Andrew R. Davis. "Indole Derivatives from the Egg Masses of Murcid Molluscs." *Molecules* 6, no. 2 (2001): 70–78.

——. "Tyrian Purple Precursors in the Egg Masses of the Australian Muricid, *Dicathais orbita*: A Possible Defensive Role." *Journal of Chemical Ecology* 26, no. 4 (2000): 1037–50.

Bijovsky, Gabriela. "The Ambrosial Rocks and the Sacred Precinct of Melqart in Tyre." In *XIII Congreso Internacional De Numismatica, Madrid, 2003: Actas-proceedings-actes,* edited by C. Alfaro, C. Marcos, and P. Otero, 829–34. Madrid: Ministerio de Cultura, Secretaría General Técnica, 2005.

"BioBriefs." *BioScience* 35, no. 8 (1985): 527. doi: 10.2307/1309832.

Bizio, Bartolemeo. "Scoperta Del Principio Purpureo Nei Due Murex Brandaris E Trunculus Linn, E Studio Delle Sue Propriet?" *Annali Delle Scienze Del Regno Lombardo-Veneto,* 1833, 346–64.

Blegen, C. W. "Excavations at Troy." *American Journal of Archaeology* 61 (1937): 582.

Blum, Hartmut. *Purpur Als Statussymbol in Der Griechischen Welt.* Bonn: R. Habelt, 1998.

Bochart, S. *Hierozoicon.* London, 1663.

Born, W. "Purple in Classical Antiquity." *Ciba Review* 1, no. 4 (1937): 111–18.

——. "Purple in the Middle Ages." *Ciba Review* 1, no. 4 (1937): 119–23.

——. "Purpura Shell-Fish." *Ciba Review* 1, no. 4 (1937): 106–10.

—. "The Use of Purple among the Indians of Central America." *Ciba Review* 1, no. 4 (1937): 124–27.

Bouchilloux, S., and J. Roche. "Sur La Pourpre Des Murex Trunculus Et Ses Precurseurs." *Comptes Rendus Des Seances De La Societe De Biologie* 148 (1954): 1583–87.

Bourrienne, Louis Antoine Fauvelet de, and Ramsay Weston Phipps. *Memoirs of Napoleon Bonaparte.* John Boyd Thacher Collection (Library of Congress). New York: Charles Scribner's Sons, 1891.

Brenner, Athalya. *Colour Terms in the Old Testament.* Sheffield: JSOT Press, Dept. of Biblical Studies, University of Sheffield, 1982.

Brett, Michael. "Carthage: The God in the Stone." *History Today* 47, no. 2 (February 1997): 44–50.

Bridgeman, Jane. "Purple Dye in Late Antiquity and Byzantium." In *The Royal Purple and the Biblical Blue: Argaman and Tekhelet: The Study of Chief Rabbi Dr. Isaac Herzog on the Dye Industries in Ancient Israel and Recent Scientific Contributions,* by Herzog and Spanier, 147–58. Jerusalem: Keter, 1987.

Brody, Aaron. "From the Hills of Adonis through the Pillars of Hercules: Recent Advances in the Archaeology of Canaan and Phoenicia." *Near Eastern Archaeology* 65, no. 1 (2002): 69–80.

Brooks, Michael. *13 Things That Don't Make Sense: The Most Baffling Scientific Mysteries of Our Time.* New York: Doubleday, 2008.

Brown, Richard. *Domestic Architecture: Containing a History of the Science, and the Principles of Designing Public Buildings, Private Dwelling-Houses, Country Mansions, and Suburban Villas, from the Choice of the Spot to the Completion of the Appendages, with Observations on Rural Residences, Their Situation and Scenery, and*

Instructions on the Art of Laying out and Embellishing Grounds.
London: G. Virtue, 1841.

Bruin, Frans. "Royal Purple and the Dye Industries of the My-
caeans and Phoenicians." *American University of Beirut Festival
Book,* 1966, 295–325.

Brunello, F. *The Art of Dyeing in the History of Mankind.* Venice:
Nerioi Pozza, 1973.

Buber, Martin. *Der Grosse Maggid und Seine Nachfolge.* Berlin: Im
Schocken, 1937.

——. *Tales of the Hasidim, the Later Masters.* New York: Schocken
Books, 1970.

Burshtein, Menachem. *HaTekhelet.* Tel Aviv: Sifriyati, 1987.

Carriker, M. R. "Comparative Functional Morphology of Boring
Mechanisms in Gastropods." *Am. Zoologist* 1 (1961): 263–65.

——. "Shell Penetration and Feeding by Naticacean and Mu-
ricacean Predatory Gastropods: A Synthesis." *Malacologia* 20,
no. 2 (1981): 403–22.

Carter, Robert, and Robert Killick, eds. *Al-Khor Island Investigating
Coastal Exploitation in Bronze Age Qatar.* Ludlow, UK: Moon-
rise, 2010.

Charles-Picard, Gilbert, and Colette Charles-Picard. *Daily Life in
Carthage at the Time of Hannibal.* New York: Macmillan, 1961.

Chateaubriand, François-René de. "Chateaubriand's Memoirs."
Wikilivres. Accessed March 28, 2012.
http://wikilivres.info/wiki/Chateaubriand's_memoirs.
Translation, A. S. Kline, book 19, chapter 16.

Chizik, Baruch. *Otzar HaTzemachim,* Volume 22, No 225, (1944):
326-345.

Christie, R. M. *Colour Chemistry.* Cambridge, UK: Royal Society of
Chemistry, 2001.

—. "Why Is Indigo Blue?" *Biotechnic & Histochemistry* 82, no. 2 (2007): 51–56. doi:10.1080/00958970701267276.

Christophersen, Carsten, F. Waetjen, O. Buchart, and U. Anthoni. "A Revised Structure of Tyreiverdin the Precursor of Tyrian Purple." *Tetrahedron* 34, no. 18 (1978): 2779–81.

Clark, Robin J. H., and C. J. Cooksey. "Bromoindirubins: The Synthesis and Properties of Minor Components of Tyrian Purple and the Composition of the Colorant from Nucella Lapillus." *Journal of the Society of Dyers and Colourists* 113, no. 11 (1997): 316–21.

—. "Monobromoindigos: A New General Synthesisthe Characterization of All Four Isomers and an Investigation into the Purple Colour of 6,6'-dibromoindigo." *New Journal of Chemistry,* 1999, 323–28.

Clark, Robin J. H., C. J. Cooksey, M. A. M. Daniels, and R. Withnal. "Indigo, Woad and Tyrian Purple: Important Vat Dyes from Antiquity to the Present." *Endeavour, New Series* 17, no. 4 (1993): 191–99.

Clelland, Eric S. *Vacuolar-type ATPase in the Accessory Boring Organ of Nucella Lamellosa (Mollusca: Gastropoda) Role in Shell Penetration.* Ottawa: Bibliotheque Nationale Du Canada, 2000.

Cohen, Alfred. *Tekhelet: The Renaissance of a Mitzvah.* New York: Michael Scharf Publication Trust of Yeshiva University Press, 1996.

Cole, William. "Purple Fish." *Philosophical Transactions of the Royal Society of London* 15 (1685): 1278–86.

Coleby, L. J. M. "A History of Prussian Blue." *Annals of Science* 4 (1939): 11–206.

Columbia College (Columbia University). *Introduction to Contemporary Civilization in the West.* 3rd ed. New York: Columbia

University Press, 1960.

Cook, A. H. "On the Shell Mound at Sidon." *Proc. Malac. Soc.* 8 (1909): 341.

Cooksey, Chrisopher J. "Bibliography of Tyrian Purple." Accessed March 27, 2012. http://www.chriscooksey.demon.co.uk/tyrian/cjcbiblio.html.

—. "Making Tyrian Purple." *Dyes in History and Archaeology* 13 (1994): 7–13.

—. "The Synthesis and Properties of 6-Bromoindigo: Indigo Blue or Tyrian Purple? The Effect of Physical State on the Colours of Indigo and Bromoindigos." *Dyes in History and Archaeology* 16/17 (2001): 97–104.

—. "The Synthesis of Minor Components of Shellfish Purple: Bromoisatin, Bromoindigotin and Bromoindirubins from Dibromoindigotin." *Beiträge Zur Waidtagung* 7 (1998): 71–74.

—. "Tyrian Purple: 6,6'-Dibromoindigo and Related Compounds." *Molecules* 6, no. 9 (2001): 736–69.

—. "Tyrian Purple: 6,6'-Dibromoindigo and Related Compounds." *Molecules* 6, no. 9 (2001): 736–69. doi: 10.3390/60900736.

Cooksey, Christopher J., and R. S. Sinclair. "Colour Variations in Tyrian Purple Dyeing." *Dyes in History and Archaeology* 20 (2005): 127–35.

Culham, Phylls. "Again, What Meaning Lies in Colour." *Zeitschrift Für Papyrologie Und Epigraphik* 64 (1986): 235–45.

Cvikel, Deborah, Yaacov Kahanov, Haim Goren, Elisabetta Boaretto, and Kurt Raveh. "Napoleon Bonaparte's Adventure in Tantura Lagoon: Historical and Archaeological Evidence." *Israel Exploration Journal* 58, no. 2 (2008): 199.

Danker, Fredrick William. "Purple." In *The Anchor Bible Dictionary*, vol. 5, edited by David Noel Freedman, 557–59. New York: Doubleday, 1992.

Dedekind, Alexander. *Ein Beitrag Zur Purpurkunde*. Berlin, 1898–1911.

Dendel, Esther Warner. *You Cannot Unsneeze a Sneeze and Other Tales from Liberia*. Niwot: University Press of Colorado, 1995.

Deshpande, Adwait M. "Indigo Dyeing." Bachelor's thesis, Institute of Chemical Technology, Mumbai, 2010.

Domínguez-Ojeda, Delia, "Biological aspects of snails *Plicopurpura pansa* and *Plicopurpura columellaris* through observations in laboratory conditions.", REDVET. Revista electrónica de Veterinaria. ISSN: 1695-7504 (2009) Vol. 10, N° 2.

Dothan, M. "Tel Mor." In *Encyclopedia of Archaeological Excavations in the Holy Land,* vol. 3, edited by Michael Avi-Yonah and Ephraim Stern, 889–90. Englewood Cliffs, N.J.: Prentice Hall, 1977.

Doumet, Joseph. *A Study on the Ancient Purple Colour and an Attempt to Reproduce the Dyeing Procedure of Tyre as Described by Pliny the Elder*. Beirut: Imprimerie Catholique, 1980.

Driessen, L. A. "Uber Eine Charakterische Reaktion Des Antiken Purpurs Auf Der Faser." *Melliand Textilberichte* 25 (1944): 66.

Dronsfield, Alan, and John Edmonds. *The Transition from Natural to Synthetic Dyes: 1856–1920*. Little Chalfont, UK: J. Edmonds, 2001.

Druding, Susan C. "History of Dyes from 2600 BC to 20th Century – Natural Dyes, Synthetic." Straw.com. Accessed March 19, 2012. http://www.straw.com/sig/dyehist.html.

Drummond, Keith N., Alfred F. Michael, Robert A. Ulstrom, and Robert A. Good. "The Blue Diaper Syndrome: Familial Hypercalcemia with Nephrocalcinosis and Indicanuria; A

BIBLIOGRAPHY

New Familial Disease, with Definition of the Metabolic Abnormality." *The American Journal of Medicine* 37, no. 6 (1964): 928–48.

Dubois, Raphael. "Recherches Sur Le Pourpre Et Sur Quelques Autres Pigments Animauxs." *Archives De Zoologie Experimentale Et Generale* 2 (1909): 471–590.

Edelstein, Sidney M. "Dyeing Fabrics in Sixteenth-Century Venice." *Technology and Culture* 7, no. 3 (1966): 395–97.

——. "Historical Notes on the Vat Processing Industry." New York: Dexter Chemical Corporation, 1972.

Edens, Christopher. "Khor Ile-Sud, Qatar: The Archaeology of Late Bronze Age Purple-Dye Production in the Arabian Gulf." *Iraq* 61 (1999): 71–88.

Edmonds, John. *The History of Woad and the Medieval Woad Vat.* Little Chalfont, UK: J. Edmonds, 2000.

——. *Medieval Textile Dyeing.* Little Chalfont, UK: J. Edmonds, 2003.

——. *The Mystery of Imperial Purple Dye.* Little Chalfont, UK: John Edmonds, 2000.

Elayi, Josette. *The Phoenician Cities in the Persian Period.* New York: ANE Society, 1981.

Eliash, Shulamit, Itamar Warhaftig, and Uri Dosberg, eds. *Maśu'ah Le-Yitshak: Sefer Zikaron Le-maran Ha-ga'on Ha-Rav Yitshak Ayzik Ha-Levi Hertsog, Zatsal, Ha-rav Ha-rashi Le-Yiśra'el, Bi-melot hamishim Shanah Li-fetirato.* Jerusalem: Yad Ha-Rav Hertsog, Mekhon Ha-Entsiklopedyah Ha-Talmudit, U-Mekhon Ha-Talmud Ha-Yiśre'eli Ha-shalem, 2008.

Elliott, Charlene. "Purple Pasts: Color Codification in the Ancient World." *Law & Social Inquiry* 33, no. 1 (2008): 173–94.

Elsner, Otto. "Solution of the Enigmas of Dyeing Tyrian Purple and the Biblical Tekhelet." *Dyes in History and Archaeology* 10

(1992): 11–16.

Elsner, Otto, and Ehud Spanier. "The Dyeing with Murex Extracts, An Unusual Dyeing Method of Wool to the Biblical Sky Blue." In *Proceedings of the 7th International Wool Textile Research Conference, Tokyo, 1985,* vol. 5, 118–30. Tokyo, Japan: Society of Fiber Science and Technology, Japan, 1985.

——. "The Dyeing with Purpura Haemastoma." Lecture presented at the 7th International Symposium on Fiber Science and Technology, Tokyo, Japan, 1985.

——. "The Past, Present and Future of Tekhelet." In Herzog and Spanier, *The Royal Purple and the Biblical Blue,* 167–77.

Epp, Dianne N. *The Chemistry of Vat Dyes.* Middletown, OH: Terrific Science, 1995.

Evans, Arthur, and Joan Evans. *The Palace of Minos; a Comparative Account of the Successive Stages of the Early Cretan Civilization as Illustrated by the Discoveries at Knossos.* London: Macmillan, 1921.

Faber, G. A. *Dyeing and Tanning in Classical Antiquity.* Basel, Switzerland: Society of Chemical Industry in Basle (Switzerland). 1938.

Faierstein, Morris M. *All Is in the Hands of Heaven: The Teachings of Rabbi Mordecai Joseph Leiner of Izbica.* Hoboken, NJ: Ktav, 1989.

Feliks, Yehuda. *The Animal World of the Bible.* Tel-Aviv: Sinai, 1962.

——. *Ha'Chai B'Mishna.* Jerusalem: Ha'Machon L'Cheker Ha'Mishna, 1972.

——. *Ha'Chai Shel Ha'Tanach.* Tel-Aviv: Sinai, 1954.

——. "Tekhelet V'Atzi Shitim B'Mishkan." *Teva V'Aretz B'Tanach,* 1992, 48–52.

Finlay, Robert. "Weaving the Rainbow: Visions of Color in

World History." *Journal of World History* 18, no. 4 (2007): 383–431. doi:10.1353/jwh.2008.0001.

Finlay, Victoria. *Color: A Natural History of the Palette.* New York: Ballantine Books, 2002.

Flemming, Wallace Bruce. *The History of Tyre.* New York: Columbia University Press, 1915.

Forbes, Robert J. *Studies in Ancient Technology.* Vol. 3. Leiden, The Netherlands: Brill, 1965.

Fouquet, Herbert, and H. J. Bielig. "Biological Precursors and Genesis of Tyrian Purple." *Angewandte Chemie International Edition in English* 10 (1971): 816–17.

Fox, Denis L. *Animal Biochromes and Structural Colors.* London: Cambridge University Press, 1953.

——. *Biochromy, Natural Coloration of Living Things.* Berkeley: University of California Press, 1979.

Fox, Harold Munro. *Blue Blood in Animals and Other Essays in Biology.* London: G. Routledge, 1928.

Fox, Harold Munro, and Gwynne Vevers. *The Nature of Animal Colours.* London: Sidgwick and Jackson, 1960.

Fox, Joseph. *Rabbi Menachem Mendel of Kotzk: A Biographical Study of the Chasidic Master.* New York: Bash Publications, 1988.

Friedländer, M. Paul. "Sur La Matiere Colorante De La Pourpre Antique." *Moniteur Scientifique,* 1909, 570.

——. "Ueber Den Antiken Purpur." *Angew Chemie* 22 (1909): 2321–24.

Fujise, Yutaka. "Chemistry of the Production of Tyrian Purple and Related Natural Products." *Kagakushi* 26, no. 1 (1999): 34–44.

Fujise, Yutaka, K. Miwa, and S. Ito. "Structure of Tyriverdin, the Immediate Precursor of Tyrian Purple." *Chemistry Letters* 6

(1980): 631–32.

Gage, John. *Color and Meaning: Art, Science, and Symbolism.* Berkeley: University of California Press, 1999.

—. "Color in Western Art: An Issue?" *Art Bulletin* 72, no. 4 (1990): 518–41.

Garfield, Simon. *Mauve: How One Man Invented a Color That Changed the World.* New York: W. W. Norton, 2001.

Gelbshtein, Hillel Meshel. "P'Sil Tekhelet." In *Mishkenot L'Abir Yaakov.* Jerusalem, 1893.

Gesenius, Wilhelm. *Hebrew and Chaldee Lexicon.* Grand Rapids: Wm. B. Eerdmans, 1950.

Ghiretti, F. "Bartolomeo Bizio and the Rediscovery of Tyrian Purple,"

Gilboa, Ayelet, Ilan Sharon, and Jeffrey Zorn. "Dor and Iron Age Chronology: Scarabs, Ceramic Sequence and 14C." *Journal of the Institute of Archaeology of Tel Aviv University* 1 (2004): 32–59.

Godard, Louis, Judith Lange, and Alexandra Doumas. *The Phaistos Disc: The Enigma of an Aegean Script.* [Greece]: Editions Itanos, 1995.

Goethe, Johann Wolfgang von. *Goethe's Theory Of Colours; Translated From The German: With Notes By Charles Lock Eastlake.* London: Murray, 1840.

Gordon, David G. *The Secret World of Slugs and Snails: Life in the Very Slow Lane.* Seattle: Sasquatch Books, 2010.

Gore, Rick. "Who Were the Phoenicians?" *National Geographic* 206, no. 4 (2004): 26-49.

Greenfield, Amy Butler. *A Perfect Red: Empire, Espionage, and the Quest for the Color of Desire.* New York: HarperCollins, 2005.

Grill-Spector, Kalanit. "Occipital Lobe." In *Encyclopedia of the Neurological Sciences,* edited by Michael J. Aminoff and Robert

BIBLIOGRAPHY

B. Daroff. Amsterdam: Academic Press, 2003.

Guerlac, Henry. "Can There Be Colors in the Dark? Physical Color Theory Before Newton." *Journal of the History of Ideas* 47, no. 1 (1986): 3–20.

Hadjikhani, N., AK Liu, AM Dale, P. Cavanagh, and RB Tootell. "Retinotopy and Color Sensitivity in Human Visual Cortical Area V8." *Nature Neuroscience* 1, no. 3 (1998): 235–41.

Haldane, Douglas. "Anchors of Antiquity." *The Biblical Archaeologist* 53, no. 1 (1990): 19–24.

Hathaway, Jennifer J. M., Coen M. Adema, Barbara A. Stout, Charlotte D. Mobarak, and Eric S. Loker. "Identification of Protein Components of Egg Masses Indicates Parental Investment in Immunoprotection of Offspring by Biomphalaria Glabrata (Gastropoda, Mollusca)." *Developmental & Comparative Immunology* 34, no. 4 (2010): 425–35. doi:10.1016/j.dci.2009.12.001.

Herm, Gerhard. *The Phoenicians: The Purple Empire of the Ancient World.* New York: William Morrow, 1975.

Herzberg, Gerhard. *Spectra of Diatomic Molecules.* New York: Van Nostrand, 1950.

Herzog, Chaim. *Living History: A Memoir.* London: Weidenfeld & Nicolson, 1997.

Herzog, Isaac. *The Dyeing of Purple in Ancient Israel.* Nahariya: Israel Malacological Society and the Municipal Malacological Museum Nahariya, 1981.

———. "Ha'Chillazon B'Clal." *Ha'Hed* 11 (1932): 20-21.

———. "Ha'Chillazon Shel Tekhelet Al Pi Hachakirot Ha'Archiologiot B'Siyua Nisyonot Maassim." *Ha'Hed* 12 (1934): 30-31

———. "Ha'Chillazon Shel Tekhelet B'Divrei Chazal." *Ha'Hed* 12 (1934): 17-21.

——. "Hatekhelet B'Yisrael." *Ha'Hed* 9 (1932): 20-22.

——. "Mareh Ha'Tekhelet." *Ha'Hed* 14 (1935): 19-22.

——. *P'sakim U'Chtavim.* Jerusalem: Mosad Harav Kook, 1989.

Herzog, Isaac, and Ehud Spanier. *The Royal Purple and the Biblical Blue: Argaman and Tekhelet: The Study of Chief Rabbi Dr. Isaac Herzog on the Dye Industries in Ancient Israel and Recent Scientific Contributions.* Jerusalem: Keter, 1987.

Hirschberg, A. S. *Ha'Halbasha Ha'Ivrit Ha'Keduma (Tekufat Ha'Mikra).* Warsaw: Edelstein, 1911.

"History & Future – India, Indigo and Independence." Human Touch Of Chemistry. Accessed March 29, 2012. http://www.humantouchofchemistry.com/india-indigo-and-independence.htm.

Hiyoshi, Y. "Chemical Education with Familiar Material in Our Hometown – Dyeing with Tyrian Purple from Thais Bronni." *Kagaku to Kyoiku* 37 (1989): 654–55.

Hoberman, Oren. "Why Is There No Blue in the Bible?" Accessed March 29, 2012. http://www.calcalist.co.il/articles/0,7340,L-3524269,00.html.

Hoeppe, Götz. *Why the Sky Is Blue: Discovering the Color of Life.* Translated by John Stewart. Princeton, NJ: Princeton University Press, 2007.

Hoffman, Rina C., Reut C. Zilber, and Roy E. Hoffman. "NMR Spectroscopic Study of the Murex Trunculus Dyeing Process." *Magn. Reson. Chem. Magnetic Resonance in Chemistry* 48, no. 11 (2010): 892–95.

Hoffmann, Roald. "Marginalia – Blue as the Sea." *American Scientist* 78, no. 4 (1990): 308–9.

Hoffmann, Roald, and Shira Leibowitz Schmidt. *Old Wine, New Flasks: Reflections on Science and Jewish Tradition.* New York:

W.H. Freeman, 1997.

Hoffman, Roy E. "The Identity of Teckelet (Biblical Blue Dye): New Findings." To be printed in *Bekhol Derakhekha Daehu: Journal of Torah and Scholarship*. Ramat Gan: Bar Ilan University Press.

Hoffner, Harry A. "Ugaritic Pwt: A Term from the Early Canaanite Dyeing Industry." *Journal of the American Oriental Society* 87, no. 3 (1967): 300-303.

Horan, P. *Ha'Tekstil B'Yemay Ha'Mikra V'Ha'Talmud*. Tel Aviv: Hamachon Lepiryon Havodah Vehayetzur, 1970.

Horowitz, Wayne, Takayoshi Oshima, and Seth L. Sanders. *Cuneiform in Canaan: Cuneiform Sources from the Land of Israel in Ancient Times*. Jerusalem: Israel Exploration Society, 2006.

Hubschman, Jerry H. "The Lowly Inertebrates: An Historical Perspective." *The Ohio Journal of Science* 79, no. 6 (1979): 243–48.

Huxley, J. "Lebanon: Phoenician Land." In *From an Antique Land*, 73–76. New York: Crown, 1954.

Hyde, Nina. "Wool – Fabric of History." *National Geographic*, 1988, 552–89.

Hyman, John. *The Objective Eye: Color, Form, and Reality in the Theory of Art*. Chicago: University of Chicago Press, 2006.

James, M. A., A. J. Mukherjee, F. Robertson, R. P. Evershed, N. Reifarth, M. P. Crump, P. J. Gates, P. Sandor, and P. Pfalzner. "High Prestige Royal Purple Dyed Textiles from the Bronze Age Royal Tomb at Qatna, Syria." *Antiquity* 83, no. 322 (2009): 1109–18.

Jensen, Lloyd B. "The Royal Purple of Tyre." *Journal of Near Eastern Studies* 22 (1963): 104–18.

Kandinsky, Wassily. *On the Spiritual in Art*. New York: Solomon

R. Guggenheim Foundation, for the Museum of Non-Objective Painting, 1946.

Kantor, Israel. "Yisrael Kotzker Goes with the Rebbe to Find the Snail." JewishGen. Accessed March 19, 2012. http://www.jewishgen.org/yizkor/radzyn/rad113.html.

Kardara, Chrysoula. "Dyeing and Weaving Works at Isthmia." *American Journal of Archaeology* 65, no. 3 (1961): 261–66.

Karmon, Nira. "The Purple Dye Industry in Antiquity." In *Colors from Nature: Natural Colors in Ancient Times,* edited by C. Sorek and E. Ayalon, 35–37. Tel Aviv: Eretz Israel Museum, 1993.

——. *Taasiat Haargaman B'Eit Ha'Atika B'Agan Ha'Mizrachi Shel Hayam Hatichon.* Haifa, Israel: Hafakulta L'Maddai Haruach, Haifa University, 1986.

Karmon, Nira, and Ehud Spanier. "Archaeological Evidence of the Purple Dye Industry from Israel." In Herzog and Spanier, *The Royal Purple and the Biblical Blue,* 147–58.

——. "Remains of a Purple Dye Industry Found at Tel Shiqmona." *Israel Exploration Journal* 38 (1988): 184–87.

Karmous, T., A. Alatrache, and N. Ayed. "Chemistry as a Tool for Differentiation Between Natural and Synthetic Dyes." *Bulletin – Union Des Physiciens* 94, no. 820 (2000): 13–23.

Kenrick, J. "Note on the Natural History of the Buccinum and Murex." In *Phoenicia,* chap. 8, 237–59. London: B. Fellowes, 1855.

Khalifeh, Issam A. *Sarepta II: The Late Bronze and Iron Age Periods of Area II, X: The University Museum of the University of Pennsylvania Excavations at Sarafand, Lebanon.* Beirut: Distribution, Departement Des Publications De L'Universite Libanaise, 1988.

Kim, SA, YC Kim, SW Kim, SH Lee, JJ Min, SG Ahn, and JH

BIBLIOGRAPHY

Yoon. "Antitumor Activity of Novel Indirubin Derivatives in Rat Tumor Model." *Clinical Cancer Research: An Official Journal of the American Association for Cancer Research* 13, no. 1 (2007): 253–59.

Kinoshita, Shūichi. *Structural Colors in the Realm of Nature.* Singapore: World Scientific, 2008.

Kitrosski, Lev. "Success of Science and Religion." *Okna,* 1994, 18.

Klein, Yves. "Yves Klein Archives." Yves Klein Archives. Accessed May 29, 2012. http://www.yveskleinarchives.org/documents/bio_us.html.

Kolb, Frank. "Review of History of Purple as a Status Symbol in Antiquity by Meyer Reinhold." *Gnomon* 45, no. 1 (1973): 50–58.

Koren, Zvi C. "Archaeo-chemical Analysis of Royal Purple on a Darius I Stone Jar." *Microchimica Acta* 162, no. 3–4 (2008): 381–92. doi:10.1007/s00604-007-0862-4.

———. "Color My World: A Personal Scientific Odyssey into the Art of Ancient Dyes." In *For the Sake of Humanity: Essays in Honour of Clemens N. Nathan,* edited by Clemens N. Nathan, Alan Stephens, and Raphael Walden, 155–89. Leiden, The Netherlands: Nijhoff, 2006.

———. "The Colors and Dyes on Ancient Textiles in Israel." In *Colors from Nature: Natural Colors in Ancient Times,* edited by C. Sorek and E. Ayalon, 15–31. Tel Aviv: Eretz Israel Museum, 1993.

———. "An Efficient High-Performance Liquid Chromatographic Analysis Scheme for Plant and Animal Red, Blue and Purple Dyes." *Dyes in History and Archaeology* 13 (1995): 27–37.

———. "High-Performance Liquid Chromatographic Analysis of an Ancient Tyrian Purple Dyeing Vat from Israel." *Israel Journal of Chemistry* 35 (1995): 117–24.

———. "Methods of Dye Analysis Used at the Shenkar College Edelstein Center in Israel." *Dyes in History and Archaeology* 11 (1993): 25–33.

Kramer, Samuel Noah. *The Sumerians: Their History, Culture, and Character.* Chicago: University of Chicago Press, 1963.

Kugel, James L. *In the Valley of the Shadow: On the Foundations of Religious Belief (and Their Connection to a Certain, Fleeting State of Mind).* New York: Free Press, 2011.

Kurlansky, Mark. *Salt: A World History.* New York: Walker and Co, 2002.

Lacaze-Duthiers, Henri. "Mémoire Sur La Pourpre." *Annales Des Sciences Naturelles,* 4th ser. *Zoologie* 12 (1859): 5–84.

Lagowski, Joseph J., ed. "Chemistry of Vision." In *Macmillan Encyclopedia of Chemistry,* vol. 4, 1492–95. New York: Macmillan Reference, 1997.

Lainer, Hayim Simhah, and Yeruham Lainer. *Sefer Dor Yesharim. Ma'amar Zikaron La-rishonim: Toldot Ha-admurim Me-Izbitsa-Radzin Z.l.l.h.h.* Jerusalem: Y. Lainer, M. Lainer, 1997.

Lamberg-Karlovsky, C. C., and P. R. S. Moorey. "Our Past Matters: Materials and Industries of the Ancient Near East." *Journal of the American Oriental Society* 117, no. 1 (1997): 87–102.

Lanigan, Leonard Ralph. "The Purple Dye Industry at Tel Dor." Master's thesis, California State University, Sacramento, 1982.

Larsen, S., and F. Watjen. "The Crystal and Molecular Structures of Tyrian Purple (6,6'-dibromoindigotin) and 2,2'-dimethoxyindigotin." *Acta Chemica Scandanavica, Ser. A,* A34(3) (1980): 171–76.

Leclerc, S., M. Garnier, R. Hoessel, D. Marko, JA Bibb, GL Snyder, P. Greengard, J. Biernat, YZ Wu, EM Mandelkow, G. Eisenbrand, and L. Meijer. "Indirubins Inhibit Glycogen Synthase Kinase-3 Beta and CDK5/p25, Two Protein Kinases Involved in Abnormal Tau Phosphorylation in Alzheimer's Disease. A Property Common to Most Cyclin-dependent Kinase Inhibitors?" *The Journal of Biological Chemistry* 276, no. 1 (2001): 251–60.

Leiner, Gershon Henokh. *Ein HaTekhelet*. New York: Safragraph, 1954.

———. *Ptil Tekhelet*. Warsaw, 1888.

———. *Sefunei Temunei Chol*. New York: Safragraph, 1952.

Leiner, Yaakov. "Mifal Tzviat HaTekhelet BMidinat Yisrael." In *HaTekhelet BTzitzit BYameinu*, 59–63. Bnei Brak, Israel: Vaad Chasidei Radzyn Byisrael, 1954.

Leonard, Albert. "Archaeological Sources for the History of Palestine: The Late Bronze Age." *The Biblical Archaeologist* 52, no. 1 (1989): 4–39.

Levey, Martin, ed. *Archaeological Chemistry: A Symposium*. Philadelphia: University of Pennsylvania Press, 1967.

———. *Chemistry and Chemical Technology in Ancient Mesopotamia*. Amsterdam: Elsevier, 1959.

Levin, Henokh M. "The Identification of the Chilazon." Afterword to *Talmud Bavli Menachot*. Shottenstein ed. New York: Art Scroll, 2004.

Levin, Yehuda L. *Bet Kotzk*. Jerusalem: Mosad Harim Levin, 2010.

Levine, Ira N. *Quantum Chemistry*. Upper Saddle River, NJ: Prentice Hall, 2000.

Lipschitz, Israel. "Kupat Ha'Rochlim." *Tifferet Yisrael - Introduction to Seder Moed*. (Printed in most standard editions of the

Mishna).

Luckenbill, Daniel David. *The Annals of Sennacherib*. Chicago: University of Chicago Press, 1924.

Magid, Shaul. *Hasidism on the Margin: Reconciliation, Antinomianism, and Messianism in Izbica/Radzin Hasidism*. Madison: University of Wisconsin Press, 2003.

Maisler, B. "Archaeology in the State of Israel." *The Biblical Archaeologist* 15, no. 1 (1952): 18–24.

—. "Canaan and the Canaanites." *Bulletin of the American Schools of Oriental Research*, no. 102 (1946): 7–12.

Maqdissi, Michel Al-, Bonacossi Daniele Morandi, Alessandro Canci, and Marta Luciani. *The Metropolis of the Orontes*. Damascus, 2005.

Matveev, Yoel. "The Rebbe of Sinn Féin." *The Jewish Daily Forward*. Accessed March 29, 2012.

http://forward.com/articles/135979/the-rebbe-of-sinn-fein/.

McCullough, David G. *Truman*. New York: Simon & Schuster, 1992.

McGovern, Patrick E. "The Royal Purple and the Biblical Blue (Argaman and Tekhelet) – The Study of Chief Rabbi Herzog, Isaac on the Dye Industries in Ancient-Israel and Recent Scientific Contributions – Spanier, E." *Isis* 81, no. 308 (1990): 563–65.

McGovern, Patrick E., J. Lazar, and Rudolph. H. Michel. "Caveats on the Analysis of Indigoid Dyes by Mass Spectrometry." *Journal of the Society of Dyers and Colourists* 107, no. 7–8 (1991): 280–81.

McGovern, Patrick E., R. H. Michel, M. Saltzman, I. I. Ziderman, and O. Elsner. "Has Authentic Tekhelet Been Identified? Short Notes." *Basor* 269 (1988): 81–90.

BIBLIOGRAPHY

McGovern, Patrick E., and Rudolph H. Michel. "Royal Purple and the Pre-Phoenician Dye Industry of Lebanon." *Museum Applied Science Center for Archaeology Journal* 3, no. 3 (1984): 67–70.

——. "Royal Purple Dye: Its Identification by Complementary Physicochemical Techniques." *Chemtracts: Inorg. Chem.* 3, no. 1 (1991): 69–76.

——. "Royal Purple Dye: The Chemical Reconstruction of the Ancient Mediterranean Industry." *Accounts of Chemical Research* 23, no. 5 (1990): 152–58.

——. "Royal Purple Dye: Tracing Chemical Origins of the Industry." *Analytical Chemistry* 57, no. 14 (1985): 1514A–522A.

Medved, David. *Hidden Light: Science Secrets of the Bible.* New Milford, CT: Maggid Books, 2010.

Mercer, Samuel A. B. *The Tell El-Amarna Tablets.* Toronto: Macmillan Company of Canada, 1939.

Michel, Rudolph H., J. Lazar, and Patrick. E. McGovern. "The Analysis of Indigoid Dyes by Mass Spectrometry." *Journal of the Society of Dyers and Colourists* 106, no. 1 (1990): 22–25.

——. "The Chemical Composition of the Indigoid Dyes Derived from the Hypobranchial Glandular Secretions of Murex Molluscs." *Journal of the Society of Dyers and Colourists* 108, no. 3 (1992): 145–50.

——. "Indigoid Dyes in Peruvian and Coptic Textiles." *Archeomaterials* 6 (1992): 69–83.

Michel, Rudolph H., and Patrick E. McGovern. "The Chemical Processing of Royal Purple Dye: Ancient Descriptions as Elucidated by Modern Science." *Archeomaterials* 1, no. 2 (1987): 135–43.

——. "The Chemical Processing of Royal Purple Dye: Ancient Descriptions as Elucidated by Modern Science, Part II." *Archeomaterials* 4, no. 1 (1990): 97–104.

Mienis, Henk, and Ehud Spanier. "A Review of the Family Janthinidae (Mollusca Gastropoda) in Connection with the Tekhelet Dye." In Herzog and Spanier, *The Royal Purple and the Biblical Blue,* 147–58.

Milgrom, Jacob. "Of Hems and Tassels." *Biblical Archaeology Review* 9, no. 3 (1983): 61–75.

——. *The Tassel and the Tallith.* [Cincinnati]: Judaic Studies Program, University of Cincinnati, 1981.

Mirsky, Aharon, *Piyutei Yossi ben Yossi*, Jerusalem: Mosad Bialik, 1991

Moore, Hillary B. "The Biology of Purpura Lapillus. Part II: Growth. Part III: Life History and Relation to Environmental Factors." *J. Marine Biol. Assoc.* 23, no. 1 (1938): 57–74.

Moorey, P. R. S. *Ancient Mesopotamian Materials and Undustries.* Oxford: Clarendon Press, 1994.

Morgenstern, Julian. "A Chapter in the History of the High-Priesthood." *The American Journal of Semitic Languages and Literatures* 55, no. 1 (1938): 1–24.

"Moshe Cotel: A Rabbi at the Piano." The Juilliard School. Accessed March 29, 2012.
   http://www.juilliard.edu/alumni/news/spotlight/archive/2008-09/200809.php.

Murphy, Brian. *The Root of Wild Madder: Chasing the History, Mystery, and Lore of the Persian Carpet.* New York: Simon & Schuster, 2005.

Nabokov, Vladimir Vladimirovich. *Laughter in the Dark.* New York: New Directions, 1960.

BIBLIOGRAPHY

Naegel, Ludwig C. A., and J. I. M. Alvarez. "Biological and Chemical Properties of the Secretion from the Hypobranchial Gland of the Purple Snail Plicopurpura Pansa (Gould, 1853)." *Journal of Shellfish Research,* August 1, 2005, 1–17.

Nam, Sangkil, Ralf Buettner, James Turkson, Donghwa Kim, Jin Q. Cheng, Stephan Muehlbeyer, Frankie Hippe, et al. "Indirubin Derivatives Inhibit Stat3 Signaling and Induce Apoptosis in Human Cancer Cells." *Proceedings of the National Academy of Sciences of the United States of America* 102, no. 17 (2005): 5998–6003.

Naor, Bezalel. "Hearot Shonot B'Inyan Ha'Chillazon V'HaTekhelet." *Ohr Ha'Mizrach* 2 (1973): 93-97.

——. *Ba-yam derech: in the sea - a way ; pathways in the Talmud.* Yerushalayim: B. Na'or, 1983.

——. "Substituting Synthetic Dye for Hilazon: The Renewal of Techelet." *Halacha and Contemporary Society,* 1992, 107–97.

——. "Tekhelet Porpyrin U'Parphyrin." *Sinai* 23 (1989).

Nassau, Kurt. "Causes of Color." Webexhibits. Accessed March 27, 2012. http://www.webexhibits.org/causesofcolor/index.html.

——. *The Physics and Chemistry of Color: The Fifteen Causes of Color.* New York: Wiley, 2001.

Navon, Mois. "The 'Hillazon' and the Principle of 'Muttar Befikha.'" *Tora U-Madda Journal* 10 (2001): 142–62.

——. "Rav's Beautiful Ratio: An Excursion into Aesthetics." *B'Ohr Ha'Torah* 19 (2009): 77–91.

Negri, Antonio Giovanni. "Della Porpora Degli Antichi E Relazione Di Altri Lavori Eseguiti Nel Laboratorio Di Chimice Generale Della R. Universita Di Genova." *Atti Della Reale Accademia Del Lincei, 2nd Series* 3 (1875–1876): 394–442.

BIBLIOGRAPHY

Nitschke, J. L., S. R. Martin, and Y. Shalev. "Between Carmel and the Sea – Tel Dor: The Late Periods." *Near Eastern Archaeology* 74, no. 3 (2011): 132–55.

Noble, Warwick J., Rebecca R. Cocks, James O. Harris, and Kirsten Benkendorff. "Application of Anaesthetics for Sex Identification and Bioactive Compound Recovery from Wild Dicathais Orbita." *Journal of Experimental Marine Biology and Ecology Journal of Experimental Marine Biology and Ecology* 380, no. 1–2 (2009): 53–60.

Orchin, Milton. "Homogeneous Catalysis: A Wedding of Theory and Experiment. The Eugene J. Houdry Award Address." *Catalysis Reviews* 26, no. 1 (1984): 59–79. doi:10.1080/01614948408078060.

Padden, AN, VM Dillon, J. Edmonds, MD Collins, N. Alvarez, and P. John. "An Indigo-reducing Moderate Thermophile from a Woad Vat, Clostridium Isatidis Sp. Nov." *International Journal of Systematic Bacteriology* 49 (1999): 1025–31.

Pamphilius, Eusebius. *The Life of the Blessed Emperor Constantine: From AD 306 to AD 337*. Merchantville, NJ: Evolution, 2009.

Pastoureau, Michel. *Blue: The History of a Color*. Princeton: Princeton University Press, 2001.

Paul, Valerie J., K. E. Arthur, R. Ritson-Williams, C. Ross, and C. Sharp. "Chemical Defenses: From Compounds to Communities." *The Biological Bulletin* 213, no. 3 (2007): 226–51.

Pfeiffer, Robert H., and E. A. Speiser. *One Hundred New Selected Nuzi Texts*. New Haven, CT: American Schools of Oriental Research, 1936.

Pharr, Clyde. *The Theodosian Code and Novels, and the Sirmondian Constitutions*. Princeton: Princeton University Press, 1952.

Polosmak, Natalia, L.P. Kundo, G. G. Balakina et al. *Textiles From the "Frozen" Tombs in Gorny Altai 400-300 BC – An Integrate Study.* Novosibirsk: Publishing House of the Siberian branch of the Russian Academy of Sciences, 2006

Pritchard, J. B. *Recovering Sarepta, a Phoenician City.* New Jersey: Princeton University Press Princeton, 1978.

Raban, Avner. "Some Archaeological Evidence for Ancient Maritine Activities at Dor." *Sefunim* 6 (1981): 20–21.

Rabinovitz, Mordechai. "Tekhelet Me'Iyey Elisha." *Otzar Hasafrut* 2 (1889): 1-26.

Rabinowitz, L. I. *Torah and Flora, Fraudulent Flora.* New York: Sanhedrin Press, 1977.

Radwin, George E., Anthony D'Attilio, and David K. Mulliner. *Murex Shells of the World: An Illustrated Guide to the Muricidae.* Stanford, CA: Stanford University Press, 1976.

Rainey, Anson F. "Who Is a Canaanite? A Review of the Textual Evidence." *Bulletin of the American Schools of Oriental Research,* no. 304 (1996): 1–15.

Raveh, Kurt. "From Holland to the Holy Land: A Personal Quest for Napoleon." Ravehholland. Accessed March 29, 2012. http://www.napoleonicsociety.com/english/ravehholland.html.

Reaumur, R. A. F. de. "Decouverte D'une Nouvelle Teinture De Poupre." *Mem. De L'Acad. Royale Des Sciences,* 1711, 216–58.

Reese, David S. "Industrial Exploitation of Murex Shells; Purpledye and Lime Production at Sidi Khrebish, Benghazi (Berenice)." *Libyan Studies,* 1980, 79–93.

——. "Iron Age Shell Purple-dye Production in the Aegean." In *Kommos,* vol. 14, edited by Joseph W. Shaw and Maria C. Shaw, 643–47. Princeton: Princeton University Press, 2000.

——. "The Mediterranean Shell Purple-dye Industry." *American*

Journal of Archaeology 90, no. 2 (1986): 183.

——. "Molluscs from Archaeological Sites in Cyprus." *Fisheries Bulletin* 5:1–112.

——. "Palaikastro Shells and Bronze Age Purple-Dye Production in the Mediterranean Basin." *The Annual of the British School of Archaelogy at Athens* 82 (1987): 201–6.

——. "Shells from Sarepta (Lebanon) and East Mediterranean Purple-dye Production." *Mediterranean Archaeology and Archaeometry* 10, no. 1 (2010): 113–41.

——. "Whales and Shell Purple-dye at Motya (Western Sicily, Italy)", Oxford Journal of Archaeology. 24 no. 2 (2005): 107-114.

Reinhold, Meyer. *History of Purple as a Status Symbol in Antiquity.* Brussels: Collection Latomus 116, 1970.

——. "On Status Symbols in the Ancient World." *Classical Journal* 64, no. 7 (1969): 300–304.

——. "Usurpation of Status and Status Symbols in the Roman Empire." *Historia: Zeitschrift Fur Alte Geschichte* 20, no. 3 (1971): 275–302.

Rendsburg, Gary. "A Further Note on Purple Dyeing." *Biblical Archaeologist* 54 (1991): 121.

——. "Israel Without the Bible." In *The Hebrew Bible: New Insights and Scholarship,* edited by Frederick E. Greenspahn, 3–23. New York: New York University Press, 2008.

Richardson, Carol M. *Reclaiming Rome: Cardinals in the Fifteenth Century.* Leiden, The Netherlands: Brill, 2009.

Richter, Sandra L. *The Deuteronomistic History and the Name Theology: Lessakken Semo Sam in the Bible and the Ancient Near East.* Berlin: W. de Gruyter, 2002.

Robinson, J. P. "Tyrian Purple." *Sea Frontiers* 17/2 (1971): 76–81.

Rock, Yehuda. "Hatalat Ha'Tekhelet B'Tzitzit." *Techumin* 16 (1996): 412–32.

Rogers, Robert William. *Cuneiform Parallels to the Old Testament.* London: H. Frowde, Oxford University Press, 1912.

Ron, M. "Difference between Tyrian Purple and Hyacinthine Purple." *Yalkut Le-sivim Le-tekhnol U-le-minhal Shel Tekst* 104 (1985): 38–39.

Ruggieri, George D. "Drugs from the Sea." *Science* 194 (1976): 491–97.

Ruscillo, Deborah. "Reconstructing Murex Royal Purple and Biblical Blue in the Aegean." In *Archaeomalacology: Molluscs in Former Environments of Human Behaviour; Proceedings of the 9th Conference of the International Council of Archaeozoology, Durham, August 2002,* edited by Mayer Daniella E. Bar-Yosef, 99–106. Oxford: Oxbow Books, 2005.

Sacks, Oliver W. *The Island of the Colorblind.* New York: A.A. Knopf, 1997.

Sagiv, Gadi. "A New Perspective on the Tekhelet Controversy of the Late Nineteenth Century." Zion, 82 (2017), pp. 59-95.

—. "Dazzling Blue: Color Symbolism, Kabbalistic Myth, and the Evil Eye in Judaism." Numen, 64, Issue 2-3 (2017): 183-208.

—. "Deep Blue: Notes on the Jewish Snail Fight." Contemporary Jewry 35 (2015), 285-313.

Saltzman, Max. "Antique Controversy." *Jsdc* 103 (1987): 404.

—. "The Identification of Dyes in Arachaeological and Ethnographic Textiles." *Archaeological Chemistry* 2 (1978): 172–85.

Saltzman, Max, A. Keay, and J. Christensen. "The Identification of Colorants in Ancient Textiles." *Dyestuffs* 44 (1963): 241–50.

Sandberg, Gösta. *Indigo Textiles: Technique and History.* Asheville,

NC: Lark Books, 1989.

—. *The Red Dyes: Cochineal, Madder, and Murex Purple: A World Tour of Textile Techniques.* Asheville, NC: Lark Books, 1997.

Schaeffer, Claude F. A. *A City with Twin Temples of Dagon and Baal: Ras Shamra Yields Fresh Treasure to the Spade: New Discoveries Concerning the God Whose Temple Samson Pulled Down Upon Himself and the Philistines.* London, 1935.

Schatz, P. F. "Indigo and Tyrian Purple – In Nature and in the Lab." *Journal of Chemical Education* 78, no. 11 (2001): 1442–43.

Scheuer, Paul J. *Chemistry of Marine Natural Products.* New York: Academic Press, 1973.

—. "The Varied and Fascinating Chemistry of Marine Mollusks." *Israel Journal of Chemistry* 16 (1977): 52–56.

Schimelman, J. "The Royal Purple." *Irradians* 8, no. 8 (1982).

Schmidt, W. Adolph. "Die Purpurfaerberei Und Der Purpurhandel in Altertum." *Die Griechjschen Papyrusurkunden Der Koeningliohen Bibliothek Zu Berlin* 1 (1842): 212–96.

Schmidt-Colinet, Andreas. "The Textiles from Palmyra." *Aram* 7 (1995): 47–51.

Schunk, Edward. "On the Formation of Indigo Blue (part I)." *Philos. Mag. J. Sci.,* 4th ser., 10 (1855): 74–95.

Scott, Philippa. "Saudi Aramco World: Millennia of Murex." Saudi Aramco World: Millennia of Murex. Accessed May 15, 2012.

http://www.saudiaramcoworld.com/issue/200604/millennia.of.murex.htm.

Sethi, G. "Indirubin Enhances Tumor Necrosis Factor-induced Apoptosis through Modulation of Nuclear Factor-B Signaling Pathway." *Journal of Biological Chemistry* 281, no. 33 (2006):

23425–35. doi:10.1074/jbc.M602627200.

Shamir, Orit. *Textile in the Land of Israel From the Roman Period Till the Early Islamic Period in the Light of the Archaeological Find*, Thesis Submitted for the Degree Doctor of Philosophy, Jerusalem, 2006.

Sheffer, Avigail, and Amalia Tidhar. *Textile History* 22, no. 1 (1991): 3–46.

Shimoyama, S., and Y. Noda. "Non-destructive Three-dimensional Fluorescence Technique." *Dyes in History and Archaeology* 12 (1994): 50–61.

Shragai, S. Z. *Be-ma'ayane hasidut Izbitsa Radzin: Asif Mi-mishnatam Shel Bet Ya'akov ve-Tif'eret Yosef.* Jerusalem: Mosad Ha-Rav Kook, 1980.

——. *Bi-netive Hasidut Izbitsa-Radzin: Perakim Be-mishnat Bet Midrasham Shel Me'orot Izbitsa-Radzin.* Jerusalem: Sh. Z. Shragai, 1972.

Singer, Charles, E. J. Holmyard, and A. R. Hall. *A History of Technology: From Early Times to Fall of Ancient Empires.* Oxford: Clarendon Press, 1954.

Singer, Itamar. "Purple-Dyers in Lazpa." In *Anatolian Interfaces: Hittites, Greeks, and Their Neighbours: Proceedings of an International Conference on Cross-Cultural Interaction, September 17–19, 2004, Emory University, Atlanta, GA,* edited by Billie Jean Collins, Mary R. Bachvarova, and Ian Rutherford, 21–43. Proceedings of an International Conference on Cross-Cultural Interaction, September 17–19, 2004, Emory University, Atlanta, GA. Oxford: Oxbow Books, 2008.

——. "Takuhlinu and Haya, Two Governors in the Ugarit Letter from Tel Afek." *Tel Aviv, Journal of the Tel Aviv University, Institute of Archaeology* 10, no. 1 (1983): 3-25.

Singer, Mendel E. "Understanding the Criteria for the 'Chilazon.'" *Journal of Halacha and Contemporary Society* 42 (2001): 5-29.

Smith, Glenn S. "Human Color Vision and the Unsaturated Blue Color of the Daytime Sky." *American Journal of Physics* 73, no. 7 (2005): 590–97.

Spanier, Ehud. "Aspects of the Biology and Behaviour of the Purple Snail Murex Trunculus." *Isr. J. Zool.* 30 (1981): 106–7.

——. "Behavioral Ecology of the Marine Snail Trunculario99sis (Murex) Trunculus." *Developments in Arid Zone Ecology and Environmental Quality,* 1981, 65–70.

——. "Cannibalism in Muricid Snails as a Possible Explanation for Archaeological Findings." *Journal of Archaeological Science* 13 (1986): 463–68.

——. "A Fossil Record of Shell Boring: Possible Evidence for Sea Level Changes in the Red Sea, Estuarine." *Coastal and Shelf Science* 24 (1987): 873–79.

——. "Rediscovering Royal Purple and Biblical Blue." *Oceanus* 33, no. 1 (1990): 75.

Spanier, Ehud, E. Linder, and Nira Karmon. *Aspektim Archeologiim Historiim Shel Hafakat Ha'Argaman B'Tekufot Kedumot.* Haifa, Israel: Haifa University, Maritime Center. (1982)

Spanier, Ehud, and Nira Karmon. "Muricid Snails and the Ancient Dye Industries." In Herzog and Spanier, *The Royal Purple and the Biblical Blue,* 147–58.

——. "Notes Concerning the Predatory Behavior of the Purple Snail Murex Trunculus." *Levantina* 28 (1980): 321–23.

Spanier, Ehud, Nira Karmon, and E. Linder. "Bibliography Concerning Various Aspects of the Purple Dye." *Levantina* 37 (1982): 437–47.

Speiser, E. A. "The Name Phoinikes." *Language* 12 (1936): 121–26.

Sukenik, Naama. *Dyes in Textiles from the Early Roman Period in the Judean Desert Caves: Chemical, Historical, and Archeological Aspect*, Thesis Submitted for the Degree Doctor of Philosophy, Bar-Ilan University, Ramat Gan (in Hebrew)

Sukenik, Naama, David Iluz, Orit Shamir, Alexander Varvak, and Zohar Amar. "Purple-Dyed Textiles From Wadi Murabba'at, Historical, Archeological, and Chemical Aspects." *Archeological Textile Reviews* No. 55 (2014): 46-54

Steinhart, C. E. "Biology of the Blues: The Snails Behind the Ancient Dyes." *Journal of Chemical Education* 78, no. 11 (2001): 1444.

Stern, Ephraim. *Dor, Ruler of the Seas: Twelve Years of Excavations at the Israelite-Phoenician Harbor Town on the Carmel Coast.* Jerusalem: Israel Exploration Society, 1994.

Stern, Ephraim, and Ilan Sharon. "Tel Dor, Preliminary Report." *Israel Exploration Journal* 37 (1987): 208.

Stern, Ephraim, J. Berg, A. Gilboa, I. Sharon, and J. Zorn. "Tel Dor, 1994–1995: Preliminary Stratigraphic Report." *Israel Exploration Journal* 47, no. 1/2 (1997): 29–56.

Stevens, Ernest Jack. *Lights, Colors, Tones and Nature's Finer Forces.* San Francisco: E. J. Stevens Light, Color and Tone Research Laboratories, 1923.

Stickle, W. B., and T. W. Howey. "Effects of Tidal Fluctuations of Salinity on Homelymph Composition of the Southern Oyster Drill. Thais Haemastome." *Nov. Biol. Berl.* 33, no. 4 (1975): 309–22.

Stieglitz, Robert R. "Commodity Prices at Ugarit." *Journal of the American Oriental Society* 99, no. 1 (1979): 15–23.

—. "The Minoan Origin of Tyrian Purple." *Biblical Archaeologist* 57, no. 1 (1994): 46–54.

Strootman, Rudolph. "The Hellenistic Royal Court. Court Culture, Ceremonial and Ideology in Greece, Egypt and the Near East, 336–30 BCE." *Mnemosyne* 62, no. 1 (2009): 168.

Sturm, Charles F., Timothy A. Pearce, and Angel Valdes. *The Mollusks: A Guide to Their Study, Collection, and Preservation.* Boca Raton, FL: Universal Publishers, 2006.

Taitelbaum, Shlomoh. *Lulaot HaTekhelet.* Jerusalem: Ptil Tekhelet, 2000.

"Takiltu." In *The Assyrian Dictionary of the Oriental Institute of the University of Chicago,* vol. 18, edited by Erica Reiner, Richard I. Caplice, Dietz Otto Edzard, Brigitte Groneberg, Hermann Hunger, Burkhart Kienast, Marie-Christine Ludwig et al., 70–73. Chicago: Oriental Institute, 2006.

"Tapiragem and Feather Color Alteration on Live Parrots by the Peoples of." Accessed March 29, 2012. http://caiquesite.com/Published%20articles/tapiragem.htm.

Tavger, Eliyahu. "Bechinot Chadashot B'Inyan Tekhelet." *Moriah* 3–4 (1992): 72–87.

—. "B'Inyan Chidush Mitzvat Ha'Tekhelet." *Ha'Maayan,* 1997, 83–85.

—. *Klil Tekhelet.* Jerusalem: Ptil Tekhelet, 1993.

Zfiyah – Temple Researches, Volume 5: Tekhelet, Argaman, Tola'at Shani. Temple Institute, Jerusalem, 1996.

Tersakian, Krikor. "Murex: The Imperial Purple Dye of Tyre." Accessed March 29, 2012. http://www.ktersakian.com/2010/12/murex-imperial-purple-dye-of-tyre.html.

BIBLIOGRAPHY

Thompson, T. E. "A Marine Biologist at Pompeii A.D. 79." *Nature* 265:292–94.

Toombs, Lawrence E., G. Ernest Wright, Robert J. Bull, James F. Ross, Edward F. Campbell, Siegfried H. Horn, and Joseph A. Calloway. "The Fourth Campaign at Balatah (Shechem)." *Bulletin of the American Schools of Oriental Research*, no. 169 (1963): 1–60.

Turok, Marta, *El Caracol púrpura: una tradición milenaria en Oaxaca*Secretaría de Educación Pública, Dirección General de Culturas Populares, Programa de Artesanías y Culturas Populares, 1988.

Twerski, Chaim E. "Identifying the Chilazon." *Journal of Halacha and Contemporary Society* 34 (1997): 77–102.

Tyron, George W., Jr. *Manual of Conchology.* Philadelphia, 1880.

Van Alphen, J. "Remarks on the Action of Light on Several Substances, Most of Them Containing Halogen, in Particular Several Indigo Dyes, in a Reducing Medium." *Recl. Trav. Chim. Pays-Bas* 63, no. 5 (1944): 95–96.

Vance, Donald R. "Literary Sources for the History of Palestine and Syria: The Phnician Inscriptions." *The Biblical Archaeologist* 57, no. 1 (1994): 2–19.

Verhecken, Andre. "Experiences with Mollusc Purple." *La Conchiglia* 22 (1990): 250–52.

—. "Experiments with the Dyes from European Purple-Producing Molluscs." *Dyes in History and Archaeology* 12 (1994): 32–35.

—. "The Indole Pigments of Mollusca." *Annales De La Societe Royale Zoologique De Belgique* 119, no. 2 (1989): 181–97.

Vermeulen, Floris N. "A Sikil Interlude at Dor: An Analysis of Contrasting Opinions." Master's thesis, University of South

Africa, 2006.

Vidal, Jordi. "Ugarit and the Southern Levantine Sea-Ports." *Journal of the Economic and Social History of the Orient* 49, no. 3 (2006): 269–79.

Voss, G. "The Analysis of Indigoid Dyes as Leuco Forms by NMR Spectroscopy." *Journal of the Society of Dyers and Colourists* 116 (2000): 80–90.

Vuorema, Anne. "Reduction and Analysis Methods of Indigo." PhD diss., University of Turku, 2008.

Wells, H. "Feeding Habits of Murex." *Ecology* 39 (1958): 556–58.

Westenholz, Joan Goodnick. "Tamar, Qedesa, Qadistu, and Sacred Prostitution in Mesopotamia." *The Harvard Theological Review* 82, no. 3 (1989): 245–65.

Westley, Chantel B. "The Distribution, Biosynthetic Origin and Functional Significance of Tyrian Purple Precursors in the Australian Muricid Dicathais Orbita (Neogastropoda: Muricidae)." PhD diss., Flinders University, 2008.

Westley, Chantel, and Kirsten Benkendorff. "Sex-Specific Tyrian Purple Genesis: Precursor and Pigment Distribution in the Reproductive System of the Marine Mollusc, Dicathais Orbita." *Journal of Chemical Ecology* 34, no. 1 (2008): 44–56.

Wilbur, K. H., and C. M. Yonge, eds. *Physiology of Mollusca.* New York: Academic Press, 1964.

Wilford, John Noble. "Earliest Samples of Royal Purple Found." *New York Times,* March, 26, 1985, sec. C, p. 2.

William, Sir Robert. *Narrative of a Voyage along the Shores of the Mediterranean Wilde.* Dublin, 1840.

Williams, Trevor I., Charles Singer, Eric John Holmyard, and Alfred Rupert Hall. *A History of Technology.* Vol. 2, *The Mediterranean Civilizations and the Middle Ages c. 700 B.C. to c.*

BIBLIOGRAPHY

A.D. 1500. Oxford: Clarendon Press, 1979.

Wood, R. W. "The Purple Gold of Tut'ankhamun." *J. Egyptian Archaeology* 20 (1934): 62–65.

Wouters, Jan. "High-Performance Liquid Chromatography of Anthraquinones: Analysis of Plant and Insect Extracts and Dyed Textiles." *Studies in Conservation* 30 (1985): 119.

Wouters, Jan, and Andre Verhecken. "The Coccid Insect Dyes: HPLC and Computerized Diode-Array Analysis of Dyed Yarns." *Studies in Conservation* 34, no. 4 (November 1989): 189.

——. "Composition of Murex Dyes." *Journal of the Society of Dyers and Colourists* 108, no. 9 (1992): 404.

——. "High-Performance Liquid Chromatography of Blue and Purple Indigoid Natural Dyes." *Journal of the Society of Dyers and Colourists* 107, no. 7–8 (1991): 266–69.

Yadin, Yigael. *The Finds from the Bar-Kokhba Period in the Caves of the Letters*. Jerusalem: Israel Exploration Society, 1963.

Yamauchi, Edwin. "The Scythians: Invading Hordes from the Russian Steppes." *The Biblical Archaeologist* 46, no. 2 (1983): 90–99.

Yoder, Christine Roy. "The Woman of Substance (Eshet Hayil): A Socioeconomic Reading of Proverbs 31:10–31." *Journal of Biblical Literature* 122, no. 3 (2003): 427–47.

Yonge, C. M. "Marine Boring Organisms." *Research* 4 (1951): 162.

Zaccagnini, C. "The Merchant at Nuzi." *Iraq* 39, no. 2 (1977): 171–89.

Ziderman, Irving I. "Biblical Dyes of Animal Origin." *Chemistry in Britain* 22, no. 5 (1986): 419–21.

——. "Blue Thread of the Tzitzit: Was the Ancient Dye a Prussian

Blue or Tyrian Purple?" *Journal of the Society of Dyers and Colourists* 97, no. 8 (1981): 362–64.

—. "First Identification of Authentic Tekelet." *Bulletin of the American Schools of Oriental Research* 265 (1987): 25–33.

—. "Halakhic Aspects of Reviving the Ritual Tekhelet Dye in the Light of Modern Scientific Discoveries." In Herzog and Spanier, *The Royal Purple and the Biblical Blue,* 126–29.

—. "On the Identification of the Jewish Tekhelet Dye." *Gloria Maris* 24, no. 4 (1985): 77–80.

—. "Purple Dyes Made from Shellfish in Antiquity." *Review of Progress in Coloration* 16 (1986): 46–52.

—. "Seashells and Ancient Purple Dyeing." *Biblical Archaeologist* 53, no. 2 (1990): 98–101.

—. "The 3600 Years of Purple-Shell Dyeing: Characterization of Hyacinthine Purple (Tekhelet)." *Advances in Chemistry Series,* 1986, 187–98.

—. *Tochnit Tzviat P'Til Tekhelet B'Chilzonot-Yam L'Shimush B'Vigdei Talit Datiim.* Jerusalem: Hamachon Ha'Yisraeli L'Sivim, 1988.

—. "Tyrian Purple or Hyacinthine Purples?" *Chemical and Engineering News* 61, no. 24 (1983): 88.

Zollinger, Heinrich. *Color Chemistry: Syntheses, Properties, and Applications of Organic Dyes and Pigments.* Weinheim, Germany: VCH, 1991.

—. "Welche Farbe Hat Der Antike Purpur?" *Textilveredlung* 24, no. 6 (1989): 207–12.

Zorn, Jeffrey R., and Robert H. Brill. "Notes – Iron Age I Glass from Tel Dor, Israel." *Journal of Glass Studies* 49 (2007): 256.

Index

A

B

C

INDEX

INDEX

gastropods. *See also* murex snails, 130

Gelbshtein, Hillel, 110

gemstones, colors of, 177

Gesenius, Wilhelm, 104, 123

Gibson, Shimon, 62- 68

Gladstone, William, 193-194

Goethe, Johann von, 183

Gomez, Kathy, 159

Gospel of Mark, 8, 232

Great Wave off Kanagawa, The (Hokusai), 120

Greenspan, Ari, 200, 207

Guberman, Joel, 151, 200, 207

H

Hasidism, 99-100

Henry IV, King, 91

Heraclius, Emperor, 82

Herod the Great, King, 58-63, 69

Herodotus, 40

Herzog, Chaim, 115, 203

Herzog, Isaac Halevi, 13, 67, 112-128, 148, 197-198, 204

Herzog, Yaakov, 231

Herzog, Yitzchak (Bougie), 230

Hexaplex (trunculariopsis) trunculus. See also Murex trunculus, 131

hillazon. See also murex snails, 66, 124

as cuttlefish, 106

characteristics in the Talmud, 106

etymology of word, 28

Rabbi Herzog's definition of, 115

Radzyn search for, 105

Hiram, King, 37

Hirshberg, Yisrael, 50

History of Animals (Aristotle), 130, 136

Hofmann, August Wilhelm von, 10

Hokusai, Katsushika, 120

Homalocantha anatomica pele. See also murex snails, 131

Horowitz, Wayne, 216

HPLC (High-Performance Liquid Chromatograph), 140-142

Huatulco, Mexico, 160-164

hyacinthina (blue), 80

Hyman, John, 192

hypobranchial gland, 135

I

India

independence of, 96

indigo in, 66, 87, 90, 96

indigo, 10, 43, 66, 91, 127

in Liberian folk tale, 85

in South Carolina, 92

synthetic, 96

the molecule, 147, 180

Indigofera tinctoria. See also indigo, 66, 88-90

insects, dyes from, 25

Instruction in the Art of the Dyers (Rosetti), 125

Isatis tinctoria. See also indigo, woad, 66, 89

Israel and Israelites. *See also* Talmud, 9, 24-25, 28, 31, 58

J

Jazirat bin Ghanim. See Al Khor Island, 19

Jeremiah (prophet), 30, 41

INDEX

INDEX

murex snails. *See also* specific
species; *tekhelet*, 7, 129-132
ancient holding pens, 51
broken shells, 50
on Roman coin, 78
Murex trunculus. See also tekhelet, xi, 6,
127, 131
color of dye from, 123
Rabbi Herzog's research on,
121
Muricidae family. See murex snails,
130-131

N

Nabokov, Valdimir, 195
Nabopolassar, 38
Naor, Bezalel, 197
Napoleon Boneparte, General, 45-
46, 56, 251, 254, 272
Nassau, Kurt, 238
Natural History (Pliny), 149
Navon, Mois, 205
Near East, murex dyes in, 7, 21, 29
Nebuchadnezzar, 38-43
Necho II Pharaoh, 38
Nero, Emperor, 77
Newton, Isaac, 168-176, 192
Nimrod (biblical figure), 21

O

Oaxaca, Mexico, 159-161
*Objective Eye: Color, Form, and Reality
in the Theory of Art* (Hyman), 192
oracle bones, 184
Orna, Mary, 231

P

Palace of Knossos. *See also* Minoan
civilization, 15-19
Pastoureau, Michel, 188
Pazyryk, 42-43, 128
Perkin, William Henry, 10, 11
Phaistos Disc. *See also* Minoan
civilization, 16, 17
Phoenicians, 32-39, 48
photons. *See also* light, and color,
178-179
Picasso, Pablo, 190
Picts, 87
Pinckney, Eliza Lucas, 92
Pinker, Steven, 193
Piyamaradu, 29, 30
Planck, Max, 175
Plicopurpura pansa, 160-165
Pliny the Elder, 5, 40, 51, 76, 149,
210
Pope, Alexander, 169
porphyria, 76
Priam, King of Troy. See
Piyamaradu, 29
Prussian blue, 118-121, 177, 231
Ptil Tekhelet, 203
purple. *See also argamman*, 4, 7-13,
19, 127
ancient residue of, 50
restrictions on use, 79-80

Q

Qatar, dyeing in, 19
Qatna, kingdom of. See Tell el-
Mishrife, 23
quinine, 10-11

INDEX

INDEX

About the Author

Baruch Sterman, Ph.D. is a physicist who helped develop the modern techniques for dyeing tekhelet authentically. Co-founder of Ptil Tekhelet, he has written numerous articles on the subject and is considered a world expert on snail dyeing. He has worked at communications technology companies in the United States and Israel, where he lives with his wife and coauthor, **Judy Taubes Sterman**.